Oracle Autonomous Database in Enterprise Architecture

Utilize Oracle Cloud Infrastructure Autonomous Databases for better consolidation, automation, and security

Bal Mukund Sharma

Krishnakumar KM

Rashmi Panda

Oracle Autonomous Database in Enterprise Architecture

Group Product Manager: Gebin George
Senior Editor: Kinnari Chohan
Technical Editor: Jubit Pincy
Copy Editor: Safis Editing
Project Coordinator: Prajakta Naik
Proofreader: Safis Editing
Indexer: Hemangini Bari
Production Designer: Ponraj Dhandapani
Marketing Coordinator: Sonakshi Bubbar

First published: December 2022

Production reference: 3310524

Published by Packt Publishing Ltd.
Livery Place
35 Livery Street
Birmingham
B3 2PB, UK.

ISBN 978-1-80107-224-3

www.packtpub.com

It is worth mentioning that the team at Packt Publishing has been the persistent driving force behind the successful publishing of this book.

There were several obstacles that arose during the pandemic that made the fate of this book appear gloomy. But we really appreciate how the Packt Publishing team pulled through and worked hard from the beginning to the completion of the book, extensively collaborating and coordinating with multiple stakeholders.

We would like to convey special thanks to our reviewers who have been very instrumental in getting the book to its final form, including refining the overall structure.

And last but not least, we will always be grateful to our families, who have always been the motivation of our lives.

- Bal, Krishna, and Rashmi

Contributors

About the authors

Bal Mukund Sharma is a technology specialist specializing in Oracle Cloud Infrastructure, databases, and engineered systems. He has experience across industries including Telecom, healthcare, financial services, and insurance. Bal is currently working as a senior cloud practice manager with Oracle and has worked in roles such as product manager, technical delivery manager, lead DBA, content architect, software engineer, and QA engineer. He helps customers with OCI-infrastructure implementation, networking, and SecDevOPS practices, along with technologies such as high availability, engineered systems, and autonomous databases. He has been very successful in helping several customers, such as banks and Telcos, to adopt the right database and cloud strategies and complex technology implementations across the world.

Besides this, Bal enjoys cooking for friends and family.

Krishnakumar KM is a cloud architect. He holds a master's in business administration from Anna University. He started his career in 2004 as a DBA and has experience working in industries including banking, Telecom, and financial services. He has been working with Oracle for 12+ years. He is passionate about innovating, deploying, debugging, and designing cloud solutions for customers. He has extensive knowledge of computing, networks, storage, and virtualization theory and architecture. He is a part of beta testing database products and has written Oracle knowledge-base articles. He has delivered presentations through various channels, including Oracle OpenWorld. He actively participates in Oracle-related forums, such as OTN communities. He has also co-authored books on Oracle database upgrade and migration methods and Oracle database high availability.

Rashmi Panda is a database enthusiast with a keen interest in Oracle Database technologies. He has spent around 18 years in different roles, including as a developer and database administrator and consulting roles. He enjoys working with customers, providing guidance in adopting better deployment solutions to address their business transformation needs in their growth roadmap. Besides databases, he has garnered excellent experience in the data integration arena, which involves performing PoC and demos on OGG, ODI, and EDQ to meet the extensive and complex data integration needs of customers. Whether on-premises, cloud, or a hybrid kind of deployment, he has been successful in creating customer solutions in designated platforms that adhere to the customer's data security requirements.

About the reviewers

Venkata Ravi Kumar, Yenugula (YVR) is an Oracle ACE Director and **Oracle Certified Master** (**OCM**) with 25 years of experience in the **banking, financial services, and insurance** (**BFSI**) verticals. He has worked as a vice president (DBA), senior database architect, senior specialist production DBA, and Oracle engineered systems architect.

He is an **Oracle Certified Professional** (**OCP**) from Oracle 8i/9i/10g/11g/12c/19c and also an **Oracle Certified Expert** (**OCE**) in Oracle GoldenGate, RAC, Performance Tuning, Oracle Cloud Infrastructure, Terraform, and Oracle Engineered Systems (Exadata, ZDLRA, and ODA), as well as Oracle Security and **Maximum Availability Architecture** (**MAA**) certified.

He has published over 100 Oracle technology articles, including on **Oracle Technology Network** (**OTN**), OraWorld Magazine, UKOUG, OTech Magazine, and Redgate. He has spoken twice at **Oracle Open World** (**OOW**), San Francisco, US.

He has designed, architected, and implemented the **core banking system** (**CBS**) database for the central banks of two countries – India and Mahé, Seychelles. Oracle Corporation, US, awarded him the title Oracle ACE Director and published his profile in their Oracle ACE Program.

They also published his profile on their OCM list and in their Spotlight on Success stories.

Dipanjan Biswas has more than 24 years of experience in the IT industry. He has played technology and delivery leadership roles in large IT consulting organizations, niche solution providers, and start-ups. He has led enterprise data platform implementations for multiple Fortune 500 customers in the pharma, retail, and banking domains. He has architected an enterprise fraud and risk management system for one of the world's largest payment network providers.

He loves listening to music and reading books.

Sonali Malik is a mother, an engineer, a techie, a tech community leader, a mentor, and a chief architect with over 20 years of experience in Oracle Database technologies, cloud services, cloud consulting, and the IT industry. As a master principal cloud architect for strategic clients, Sonali specializes in Oracle Cloud Infrastructure and platform services, software delivery networks, cloud security, and Oracle products, along with other cloud platforms and on-premises technologies. Sonali has garnered substantial industry experience. In the past, she held technical and managerial positions in various organizations. She has an extensive history of customer advocacy with an understanding of how to suggest and help with making innovative technical decisions, get client consensus, and deliver commitments.

When she is not working, she spends her time pursuing hobbies such as dancing, reading books, cooking, traveling, and spending time with her family and friends. She also volunteers at different organizations that support diversity and inclusion.

Table of Contents

Part 1 – Understanding Autonomous Database in OCI

1

2

Autonomous Database Deployment Options in OCI   41

Part 2 – Migration and High Availability with Autonomous Database

3

Migration to Autonomous Database   69

4

5

6

Part 3 – Security and Compliance with Autonomous Database

7

Security Features in Autonomous Database   229

Index   265

Other Books You May Enjoy   274

Preface

Oracle **Autonomous Database** (**ADB**) is built on the world's fastest Oracle Database platform, Exadata, and is delivered on **Oracle Cloud Infrastructure** (**OCI**), the customer data center (ExaCC), and Oracle Dedicated Region Cloud. This book is a fast-paced, hands-on introduction to the most important aspects of OCI ADB.

You'll get to grips with the concepts that you need to understand to design disaster recovery using a standby database deployment for ADB. As you progress, you'll understand how you can take advantage of automatic backup and restore. The concluding chapters will cover topics such as the security aspects of databases to help you learn about managing ADB, along with exploring the features of ADB security, such as Data Safe and customer-managed keys for vaults.

By the end of this Oracle book, you'll be able to build and deploy ADB in OCI, migrate databases to ADB, comfortably set up additional high-availability features, such as Autonomous Data Guard, and understand end-to-end operations with ADB.

Who this book is for

ADB is based on the core Oracle Database technology, and hence it can be easily used by all users that use Oracle Database without requiring any add-on skills.

Besides users of Oracle Database, this book is a good guide for non-Oracle Database users, such as users of MS SQL or PostgreSQL, who can easily get started without going deeper into the provisioning and managing concepts of a base Oracle Database.

It's not just DBAs who are the primary audience. Instead, this book can prove beneficial for enterprise architects, IT technicians, and the broader developer community who are key decision-makers in the enterprise deployment strategy.

What this book covers

Chapter 1, *Introduction to Oracle's Autonomous Databases*, introduces the basics of the ADB model. It talks about the building blocks of ADB, its hardware details, and best practices. This chapter also emphasizes the role of ADB in real-time environments. It also discusses a few real-time use cases to demonstrate how businesses can benefit from adopting ADB. Overall, it's a very good chapter to kick off the autonomous journey.

Chapter 2, Autonomous Database Deployment Options in OCI, explains all the available options to provision ADB in OCI, including the free tier. It discusses in detail the dedicated and shared environments and their differences. Also, it provides an overview of important OCI attributes, such as IAM, networking, and security, related to ADB.

Chapter 3, Migration to Autonomous Database, helps simplify your journey of migrating to ADB in OCI. It discusses the most recommended methods to ease your migration to ADB. Different deployment options are available with ADB, but in this chapter, we focus on how you can entirely automate the migration of your Oracle databases, which may be on-premises or in a non-Oracle cloud, into ADB in shared infrastructure in the OCI public cloud. The steps remain almost the same for migration to ADB in a dedicated infrastructure or a dedicated region.

Chapter 4, ADB Disaster Protection with Autonomous Data Guard, details how you can ensure the extreme availability of ADB. In this chapter, you will understand how even disaster recovery can be fully automated and managed in an autonomous environment by just enabling Autonomous Data Guard. You will learn how to enable a local or remote standby and about the database peer states, the different operations, and the associated RPO/RTO during a failover operation.

Chapter 5, Backup and Restore with Autonomous Database in OCI, explains all the available database backup and restore options in ADB. This chapter illustrates how database backups are done manually in ADB. Also, it discusses the data replication within ADB through database links.

Chapter 6, Managing Autonomous Databases, discusses all the aspects of managing ADB. As most of the tasks are automated in ADB, this chapter gets more weightage. This chapter explains how we can effectively manage ADB through best practices and other given provisions. It also enlightens us about the tools available exclusively for ADB and its importance in database management. The topics explained in this chapter will be helpful to handle day-to-day database operations.

Chapter 7, Security Features in Autonomous Database, describes the important concepts related to database security. This chapter explains how ADB ensures security. It describes all the database security tools bundled with ADB and also clarifies the role of the user in ensuring database security.

To get the most out of this book

Software/hardware covered in the book	Operating system requirements
Oracle Cloud Infrastructure (OCI) free tier/pay-as-you-go autonomous database (ATP or ADB)	Windows, macOS, or Linux
OCI account with privileges to create an Object Storage Bucket and an autonomous database	

If you are using the digital version of this book, we advise you to type the code yourself or access the code from the book's GitHub repository (a link is available in the next section). Doing so will help you avoid any potential errors related to the copying and pasting of code.

The OCI portal interface includes GUI tools to manage ADB. All the database-related activities can be done using those tools. We won't require any third-party tools to manage or operate the database.

Download the color images

We also provide a PDF file that has color images of the screenshots and diagrams used in this book. You can download it here: `https://packt.link/H7keF`.

Conventions used

There are a number of text conventions used throughout this book.

`Code in text`: Indicates code words in text, database table names, folder names, filenames, file extensions, pathnames, dummy URLs, user input, and Twitter handles. Here is an example: "Mount the downloaded `WebStorm-10*.dmg` disk image file as another disk in your system."

A block of code is set as follows:

```
BEGIN
DBMS_CLOUD.CREATE_CREDENTIAL(
credential_name => 'Credential_Name',
username => 'Cloud_UserName',
password => '<authorization_token>'
);
END;
/
```

When we wish to draw your attention to a particular part of a code block, the relevant lines or items are set in bold:

```
ALTER DATABASE PROPERTY SET DEFAULT_CREDENTIAL = 'ADMIN.
OBJSTORE_CRED';
```

Any command-line input or output is written as follows:

```
SQL>  create directory dump_dir as 'datapump_dir';
Directory created.

SQL> select directory_name , directory_path from dba_
directories;
```

Bold: Indicates a new term, an important word, or words that you see onscreen. For instance, words in menus or dialog boxes appear in **bold**. Here is an example: "Click on **Identity | Users**."

> **Tips or important notes**
> Appear like this.

Get in touch

Feedback from our readers is always welcome.

General feedback: If you have questions about any aspect of this book, email us at customercare@packtpub.com and mention the book title in the subject of your message.

Errata: Although we have taken every care to ensure the accuracy of our content, mistakes do happen. If you have found a mistake in this book, we would be grateful if you would report this to us. Please visit www.packtpub.com/support/errata and fill in the form.

Piracy: If you come across any illegal copies of our works in any form on the internet, we would be grateful if you would provide us with the location address or website name. Please contact us at copyright@packt.com with a link to the material.

If you are interested in becoming an author: If there is a topic that you have expertise in and you are interested in either writing or contributing to a book, please visit authors.packtpub.com.

Share Your Thoughts

Once you've read *Oracle Autonomous Database in Enterprise Architecture*, we'd love to hear your thoughts! Scan the QR code below to go straight to the Amazon review page for this book and share your feedback.

https://packt.link/r/1801072248

Your review is important to us and the tech community and will help us make sure we're delivering excellent quality content.

Download a free PDF copy of this book

Thanks for purchasing this book!

Do you like to read on the go but are unable to carry your print books everywhere?

Is your eBook purchase not compatible with the device of your choice?

Don't worry, now with every Packt book you get a DRM-free PDF version of that book at no cost.

Read anywhere, any place, on any device. Search, copy, and paste code from your favorite technical books directly into your application.

The perks don't stop there, you can get exclusive access to discounts, newsletters, and great free content in your inbox daily

Follow these simple steps to get the benefits:

1. Scan the QR code or visit the link below

https://packt.link/free-ebook/9781801072243

2. Submit your proof of purchase

3. That's it! We'll send your free PDF and other benefits to your email directly

Part 1 – Understanding Autonomous Database in OCI

The objective of this part is to give you a clear understanding of **Autonomous Database** (**ADB**) concepts. You will be able to clearly articulate the business benefits and technical merits of using Oracle's ADB.

After completing *Part 1* of the book, you should be able to deploy ADB based on use cases such as transaction processing versus data warehousing and Shared versus Dedicated, all based on a better understanding of networking and IAM best practices for deployment in OCI.

This part comprises the following main chapters:

- *Chapter 1, Introduction to Oracle's Autonomous Database*
- *Chapter 2, Autonomous Database Deployment Options in OCI*

1

Introduction to Oracle's Autonomous Database

This chapter is an introduction to Oracle's **Autonomous Database** (**ADB**). It explains the hardware architecture supporting this service. You will learn how it differentiates from traditional database deployments and the reasons to select ADB. We will explore various use cases for ADB along with business benefits in terms of **Total Cost of Ownership** (**TCO**)/**Return on Investment** (**ROI**), compared to traditional deployment.

With this chapter, you will build a solid foundation on **ADB,** which will be very useful later when you start architecting your application use case.

In this chapter, we will cover the following topics:

- Learning what an Autonomous Database is
- Technology building blocks – ADB
- Classification of ADB based on workload
- ADB infrastructure deployment choices – shared or dedicated?
- Understanding why to use an Autonomous Database
- Reviewing use cases for ADB
- Understanding the business benefits of using ADB
- BOM and SKUs for Autonomous Databases

By the end, you should have a clear understanding in terms of which flavor of ADB is good for your use case.

Technical requirements

For this chapter, although there are no technical requirements, if you are familiar with the **Oracle Cloud Infrastructure** (**OCI**) Console, you will be able to visualize the topics discussed.

Learning what an Autonomous Database is

Before we start learning about ADB, let's first understand what OCI is.

OCI was designed to satisfy the needs of enterprise workloads that often require high performance, security, elasticity, availability, and integrity for their critical applications. Enterprises today want to lower their cost and move from a traditional CAPEX-based model to an OPEX-based culture. At the same time, they need a rich set of cloud services and automation capabilities built using cloud-native technologies to provide a comprehensive cloud solution for customers. OCI provides services around infrastructure, data management, analytics, applications, development/DevOps, governance, and security to cater to requirements from big to small enterprises. OCI is not just limited to Oracle's data center but can also be extended to customers' data centers; an offering called **Cloud@Customer**, which runs behind the company's firewall, is available, and solves data sovereignty requirements. OCI is also available as dedicated regions for those workloads that require an in-country location or have data sovereignty requirements.

OCI's ADB is a **self-driving**, **self-securing**, and **self-repairing** fully-managed database environment available on the cloud as well as on-premises. As of right now (the cloud is all about change), four distinct workload types are available with autonomous databases: **Autonomous Transaction Processing** (**ATP**), **Autonomous Data Warehouse** (**ADW**), **Application Express** (**APEX**), and **Autonomous JSON Database** (**AJD**).

You can build data-driven apps and gain operational insight in real time without worrying about the operational aspects of a database in terms of maintenance tasks such as **backups**, **patching**, **upgrades**, and **performance tuning**. You can scale the number of CPU cores or the storage capacity of the database at any time without impacting the availability or performance of the database system. With cloud-native developments and auto, we have highlighted and discussed auto scaling in detail in other chapters. Here, we give an overview of all the scaling features OCI provides and its automated responses to your workload needs.

ADB is built upon a very solid foundation with more than three decades of technical innovations developed by Oracle, providing customers flexibility combined with **Machine Learning** (**ML**) and **Artificial Intelligence** (**AI**). Oracle manages everything for you, so you can focus on your data, development, and delivering solutions that impact your business.

ML models and algorithms run inside Oracle ADB. It brings the following advantages:

- Data stays in place
- Massive parallel execution
- Flexible model building

In addition to these benefits, ADB also supports key Oracle database features and open source programming languages:

- **Structured Query Language (SQL)**, R, or Python
- **Oracle Data Miner (ODM)**
- Oracle AutoML

With Oracle Machine Learning and Oracle ADB, users have a variety of options for building and deploying models involved in data science projects, whether they use in-database algorithms or open source Python algorithms. An autonomous database uses AI and ML to achieve a complete, automated provisioning experience, applying security, automated patch updates, continuous availability, and performance tuning based on the workload types; managing changes; and avoiding mistakes. **Oracle Machine Learning for Python (OML4Py)** in Oracle ADB upholds versatile in-dataset information investigation and arrangement utilizing local Python grammar, conjuring in-dataset calculations for model structure and scoring, and implanting the execution of client-characterized Python capacities from Python or REST APIs. Likewise, OML4Py incorporates the AutoML interface for automated calculations and component choice and hyperparameter tuning to augment model execution. On the other hand, ODM, which is an extension of Oracle SQL Developer, helps develop ML methods.

Let's discuss a bit about security in autonomous databases. We will go through the details of it in *Chapter 7, Security Features in Autonomous Database*. All data in ADB is encrypted, and users or applications need to be authenticated in order to use the database. ADB does not require any manual configuration for providing encryption – whether data is at rest or in motion, all connections use certificate-based authentication over **Secure Socket Layer (SSL)**. ADB enforces strong password complexity for all users based on Oracle Cloud Security standards. ADB provides a network **Access Control List (ACL)**, using which databases can only accept connections from allowed IP addresses and reject all other client connections. ADB also provides network access through private endpoints that help organizations implement strict security mandates to only allow connections privately from inside a **Virtual Cloud Network (VCN)**, and traffic never uses public subnet and public internet within your tenancy VCN.

Quick note

In OCI, ADW was the first offering launched in 2018. Later, ATP was added to the service portfolio of offerings at beginning of 2019. Recently, in August 2020, Oracle also added **JavaScript Object Notation (JSON)** databases to the Autonomous Database service catalog, known as AJD.

> **Quick note**
>
> You can think of a VCN as a private network set up inside an Oracle data center, which consists of several firewall rules and communication gateways. The components of a VCN are one or more subnets, **Internet Gateways (IGWs)**, **Dynamic Routing Gateways (DRGs)**, route tables, security lists, and DHCP options. When you create a VCN inside OCI, most of these components are created by default. Another thing to keep in mind is that a VCN covers a single, continuous IPv4 CIDR block of your choice. In other words, you can say that a VCN provides software-defined networking in OCI.

Always Free ADB

Always Free ADB is available through Oracle's Free Tier, which provides customers with up to two instances of ADB (Serverless/Shared) for every tenancy. Always Free ADB supports both ATP and ADW workload types. Customers can upgrade a Free database to Paid anytime.

> **Quick note**
>
> You can sign up for an Oracle Free Tier account by navigating to `https://www.oracle.com/cloud` and clicking on the **Try Oracle Cloud Free Tier** button on the right side.

The key characteristics are as follows:

- It has a fixed configuration: 1 OCPU, 20 GB of storage, and 8 GB of memory
- Up to two Always Free instances in every tenancy's home region
- Most ADB functionality available, except **Scale Up/Down**, **Storage auto scaling**, **Update License Type**, **Manual backup**, and **Restore**
- Upgrade an Always Free database to Paid anytime

Always Free ADB gets automatically stopped after 7 days of continuous inactivity. After 90 (cumulative) days of continuous inactivity, Free ADB instances are also automatically terminated. Users are notified via console UI banners for both of these events.

Always Free autonomous databases can only be created in your account's home region as shown:

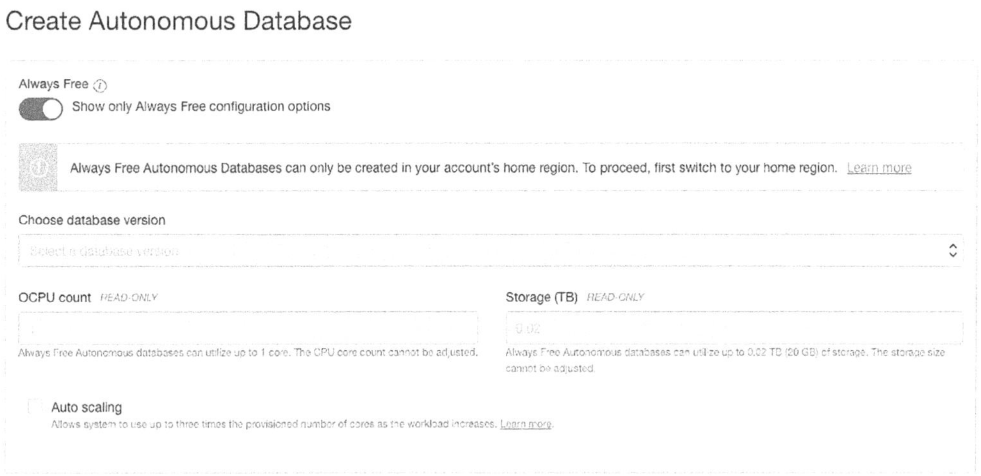

Create Autonomous Database

Always Free ⓘ

Show only Always Free configuration options

Always Free Autonomous Databases can only be created in your account's home region. To proceed, first switch to your home region. Learn more

Choose database version

Select a database version

OCPU count *READ-ONLY*

Always Free Autonomous databases can utilize up to 1 core. The CPU core count cannot be adjusted.

Storage (TB) *READ-ONLY*

0.02

Always Free Autonomous databases can utilize up to 0.02 TB (20 GB) of storage. The storage size cannot be adjusted.

Auto scaling

Allows system to use up to three times the provisioned number of cores as the workload increases. Learn more

Figure 1.1 – The Always Free ADB option during deployment

Customers can create autonomous databases quickly and easily using the OCI Console, **Command-Line Interface (CLI)**, **Software Development Kit (SDK)**, and Terraform. To create a database, customers can log in to their OCI Console and select **Oracle Database | Autonomous Data Warehouse | Transaction Processing | JSON Database**. You need to provide details such as the database name, the number of OCPUs, storage (in TB), and the admin password. In a couple of minutes, a fully ready autonomous database is ready for use by the customer. Customers can perform various management operations on their databases, such as starting, stopping, restarting, backing up, cloning, using Data Guard, and monitoring. Backups are automatic and the customer has the option to take a full backup anytime, as well as the ability to restore to a "point in time" backup. Backups are retained for 60 days by default and the customer can configure it to be more or less. The customer can scale their database CPUs and storage without any downtime. Using administration credentials, customers can access and start using the ADB service using a separate service console. You can also update admin credentials anytime.

Technically, ADB is built on OCI Exadata infrastructure. Each ADB database is an independent **Pluggable Database (PDB)** to which the customer doesn't have host access. Oracle manages the entire life cycle activities of the database based on customer inputs and preferences. You can check the ADB page within Oracle Cloud Console as depicted in the following screenshot. It shows deployed ADBs within a region.

Figure 1.2 – ADB page on the Console

As we can see in *Figure 1.2*, a single page has both a shared and dedicated infrastructure link for the easy creation of these services and navigation capabilities.

Technology building blocks – ADB

Let's see what makes an Oracle database autonomous. We will look at various building blocks for autonomous databases. You will notice that starting from Oracle Database version 9i, Oracle introduced several automation capabilities around memory management, workload monitoring, and self-tuning capabilities, which set the base for autonomous databases. With the acquisition of Sun Microsystems, Oracle drove a database infrastructure with engineered systems focused on more automation capabilities and bringing data processing to the storage layer, with innovations such as Smart Scan, query offloading, a storage index, columnar compression, and so on. These database platforms are preconfigured and highly optimized for running database workloads, pre-tested across thousands of deployments, thus forming the base for autonomous databases.

The ADB building blocks are as follows:

- Oracle **Database Enterprise Edition (DBEE)**
- Oracle Exadata Database Machine
- OCI
- ML
- Oracle's best practices
- Oracle's knowledge base

We will talk about each block in detail in the next sections.

Oracle DBEE

If you have prior knowledge of Oracle databases, you will already know that Oracle had two distinct editions of databases targeted for different market segmentation: a Standard edition and an Enterprise edition. As the Enterprise edition was built to suit the high-performance requirements of enterprise customers for transactional and analytical workloads, it has several features that make it enterprise-class. With traditional database deployments, the DBA needs to tweak several configuration parameters based on workload types, not just the database but also the **Operating System** (**OS**) and network configuration – everything that goes with any production-ready database deployment. ADB removes these complexities and comes preconfigured with optimal values based on deployment types.

Oracle DBEE sets the foundation for autonomous databases. The database options available with DBEE provide the required capabilities to run ADB. The following options give ADB autonomous capabilities:

- **Real Application Clusters** (**RAC**): Provides high availability functionality, including scale-out architecture, failover in case of instance failure, and online patching to avoid downtime

- **Active Data Guard**: Provides standby capabilities and is used for disaster recovery purposes

- **Parallel SQL**: A core feature for prioritizing SQL's parallel degree based on system resources and policies

- **Multitenant option**: Provides the required functionality for Agile development

- **Database In-Memory**: Provides high performance for analytic queries

- **Transparent Database Encryption** (**TDE**): Part of the Oracle Advanced Security options – a default for data encryption

- **Database Vault**: Segregation of duties running within a database kernel – used for compliance requirements and blacklisting and whitelisting of users and programs

> **Quick note – database parameters**
>
> All database parameters are set to optimal values based on workload type. Users can only change a limited number of parameters. These parameters are shown in the following screenshot.

Allowed Parameters for modification

CURSOR_SHARING
DDL_LOCK_TIMEOUT
FIXED_DATE
LDAP_DIRECTORY_ACCESS
MAX_IDLE_TIME
MAX_STRING_SIZE
NLS_CALENDAR
NLS_COMP
NLS_CURRENCY
NLS_DATE_FORMAT
NLS_DATE_LANGUAGE
NLS_DUAL_CURRENCY
NLS_ISO_CURRENCY
NLS_LANGUAGE
NLS_LENGTH_SEMANTICS
NLS_NCHAR_CONV_EXCP
NLS_NUMERIC_CHARACTERS
NLS_SORT
NLS_TERRITORY
NLS_TIME_FORMAT
NLS_TIME_TZ_FORMAT
NLS_TIMESTAMP_FORMAT
NLS_TIMESTAMP_TZ_FORMAT

OPTIMIZER_CAPTURE_SQL_PLAN_BASELINES (Allowed only with ALTER SESSION)
OPTIMIZER_IGNORE_HINTS
OPTIMIZER_IGNORE_PARALLEL_HINTS
OPTIMIZER_MODE
PLSCOPE_SETTINGS
PLSQL_CCFLAGS
PLSQL_DEBUG
PLSQL_OPTIMIZE_LEVEL
PLSQL_WARNINGS
QUERY_REWRITE_INTEGRITY
RESULT_CACHE_MODE
STATISTICS_LEVEL (Allowed only with ALTER SESSION)
TIME_ZONE (Allowed only with ALTER SESSION)
APPROX_FOR_AGGREGATION
APPROX_FOR_COUNT_DISTINCT
APPROX_FOR_PERCENTILE
AWR_PDB_AUTOFLUSH_ENABLED
CONTAINER_DATA
CURRENT_SCHEMA (Allowed only with ALTER SESSION)

Figure 1.3 – List of allowed parameters for modification in ADB

As we can see, most of the changeable parameters are around the user's profile, related to NLS, time zones, and so on. Oracle sets all other parameters to optimal values by default.

Oracle Exadata Database Machine

If you are new to Exadata Database Machine, you can consider it a combination of software and hardware optimized to run Oracle Database. The current version of Exadata is X9-M and it will be keep being updated based on the latest version. Normally, Oracle follows a cycle of 12 to 18 months to release a new generation of Exadata machines. The very first version of Exadata was released back in 2008. With each Exadata refresh cycle, customers get the most recent CPU processors, memory, increased disk capacity, flash, and high-speed networking components, which provide increased performance, security, availability, and management capabilities. Exadata is known as a great consolidation platform because of the massive capacity and performance available with these machines.

Oracle introduced a storage layer within the database machine, with several innovations supporting scale-out architecture, and parallel query operation, which greatly optimized data processing at its storage layer. Exadata solved two major problems: avoiding network bottlenecks for data movement within the machine through SQL offloading and, at the same time, providing a larger network bandwidth (100 Gbit/s) Ethernet fabric for data access. Exadata also provided separate Ethernet ports for data center connectivity and management operations such as backups. Some of the key innovations within machines can be considered Smart Scan, query offloading to storage, storage indexes, flash caching, resource management, **Hybrid Columnar Compression** (**HCC**), and in-memory database capabilities with fault tolerance.

Quick note – ADB platform

ADB runs on RAC on Exadata. ADB decides where to place each database during provisioning. A fewer number of instances are preferred when possible. Even though it's running on RAC, the database can only be open on one node.

OCI

OCI provides required technologies such as networking elements, VCNs, subnets, virtual firewalls (network security groups), security lists, communication gateways, identity and access control, automated provisioning, logging, audit, monitoring capabilities, and so on, which are needed to run Exadata Cloud Service natively. OCI provides end-to-end security with a focus on the unified, automated, prescriptive security experience that makes life easier for customers. Identity management is a key focus for OCI, which helps simplify a customer's security landscape, starting with data and then moving through the infrastructure, network, monitoring, and edge services. At the data or database layer, OCI supports encryption at rest and in transit and supports hardware security modules.

Within infrastructure, for compute instances, OCI supports hardened OS images, autonomous Linux, hardware root-of-trust, and signed firmware. In the networking domain, OCI supports isolated network virtualization with off-box **Network Interface Cards** (**NICs**), private networking with FastConnect, and security zones, which can be used to apply context-specific security policies to compartments. For monitoring, OCI has integrated Cloud Security Posture Management with Cloud Guard. Cloud Guard is very dynamic and OCI releases new services every week for customer needs. Recently, Oracle announced Scanning Service, which scans compute hosts and container images for vulnerabilities. The Bastion service automates the configuration of secure Bastion servers. The Certificates service automates the provisioning and management of private and public certificates. Threat Intelligence Service centralizes threat intelligence and vulnerability feeds integrated across cloud services.

Oracle best practices

Every organization emphasizes adopting best practices, Oracle has published several "best practices," which are based on expert recommendations for deploying a product, fine-tuning, configuration changes, and so on. In addition to this, Oracle's **Maximum Availability Architecture (MAA)** focuses on best practices for the availability of applications based on the categorization of **Service-Level Agreements (SLAs)**. Oracle has a set of best practices around security called defense-in-depth. Best practices allow the Oracle database to run with optimal efficiency. Oracle uses several of its core features to provide the required level of optimization; using technologies such as online reorganization allows online operation for table redefinition without compromising the availability of the system. Using Resource Manager, Flashback Technology, Application Continuity, and Transparent Application Continuity protects against several kinds of failures. RAC protects against node failures, and this also enables rolling patches, service draining, and zero-downtime planned maintenance. Application Continuity protects transactions from failures, allowing safe transaction replay using a JDBC replay driver and Transaction Guard. Using best practices ensures that Oracle technologies achieve the highest level of performance, availability, and security.

Oracle knowledge base

Oracle has a knowledge base of several years built from a diverse set of customer issues, whether service requests, product management contributions, development experiences, bugs reported by customers, or enhancement requests for features and services. The Oracle system for support tickets, known as MOS, is an interface for Oracle and customers that allows them to open up support tickets in case they need help. Customers can visit MOS to explore several knowledge base articles, how-tos, and so on. As part of problem diagnostics and resolution, customers often provide logs, screen shares, diagnostic collection, OS details, trace files, and so on. These files are a great source of information for Oracle for using these as input to build AI and ML for intelligent data analysis and problem-solving.

ML

ML is an important function in the autonomous database, where the database uses ML algorithms and automates the most important aspects of the database, such as security, database backups, patching, performance tuning, index creations, and several routine tasks that are typically performed by a DBA. It has increased productivity, as no human intervention is needed, thus freeing up time for other inventions. OCI also provides an ML platform as a service that can be used by customers to implement their own solutions while using ADB as the database of choice. ML is also a set of tools available in Oracle Cloud that customers can use to implement their own solutions.

We have learned about the different building blocks for ADB, so now it is time to look at the classification of ADB based on workload characteristics. OCI has tailored these databases based on types such as DWH, OLTP, and JSON.

Classification of ADB based on workload

With the optimization and integration of hardware and software capabilities available across the stack in database machines, along with Oracle Enterprise edition databases, Oracle offers three distinct flavors of ADB for running your workload: ATP, ADW, and AJD.

Oracle ADW is designed to run data warehousing, data marts, data lakes, analytics, and ML workloads. Oracle ATP is designed for online transaction processing, batch, reporting, the **Internet of Things (IoT)**, application development, ML, and mixed workload environments.

In the following sections, we will discuss each flavor in detail.

ADW

This is the first offering available with OCI in the ADB service portfolio. As the name indicates, Oracle ADW is designed for data warehouses and related workloads, including data marts, data lakes, and ML workloads. Most organizations architect analytical workloads to run on a separate system other than their OLTP systems, as the requirements for these systems are different and widely used for decision-making and data analytics business use cases. Data warehouses are characterized by star schemas and snowflake schemas and normally have very high data ingestion rates. As part of the data warehousing requirements, facts are often derived from several dimensions, and keeping aggregated data is often considered as summary tables for data analysis. This system demands a high level of parallelism for running SQLs as well as a faster response time to serve business users. Oracle ADW is specifically designed to provide faster response times to queries and desired level of parallel processing for data ingestion.

> **Quick note**
>
> ADW optimizes complex SQL. It uses the columnar format and creates data summaries. Optimizer and PARALLEL hints are ignored in ADW. Users can override this behavior by changing two parameters, `optimizer_ignore_hints` and `optimizer_ignore_parallel_hints`, to FALSE, which, by default, are usually set to TRUE.
>
> **Optimizer statistics**: Stats are gathered automatically for direct load operations. If your workload uses conventional DML in ADW, gather stats manually with the GATHER AUTO option. For example, see the following:
>
> ```
> BEGIN
> DBMS_STATS.GATHER_SCHEMA_STATS('SH', options=>'GATHER AUTO');
> END;
> /
> ```

ATP

Oracle ATP is designed for online transaction processing and workloads that are not data warehousing-related. ATP is primarily suited for mission-critical transactional workloads that often include operational reporting or batch data processing. With ATP, you can run mixed workloads in a single database, which eliminates the need to segregate transactional data from analytics data. Users can run their mixed workload in the same system without worrying about any potential data management options. ATP also supports the IoT and ML, in addition to OLTP workloads. ATP makes application development much simpler, as there is no need for traditional data management skills for someone to get started with these services.

> **Quick note**
>
> ATP optimizes the response time for SQLs. Data is stored in a 'ROW' format and creates indexes as required automatically. Optimizer and PARALLEL hints are honored in ATP and are set to TRUE by default. Users can override this behavior by changing two parameters, `optimizer_ignore_hints` and `optimizer_ignore_parallel_hints`, to TRUE, which are set to FALSE by default. This will disable both the default behavior and the setting.
>
> ```
> ALTER SESSION SET OPTIMIZER_IGNORE_HINTS=TRUE;
> ```
>
> ```
> ALTER SESSION SET OPTIMIZER_IGNORE_PARALLEL_HINTS=TRUE;
> ```
>
> **Optimizer statistics**: ATP gathers stats with a nightly auto stats job. Real-time statistics collection gathers a subset of optimizer statistics for conventional DML operations: number of rows, MAX and MIN column values, and so on. High-frequency statistics collection gathers full optimizer statistics every 15 minutes if statistics are stale.

AJD

Oracle AJD is a new cloud service launched by Oracle around mid-August 2020. It is built for organizations and developers who want to build interactive applications and microservices that primarily deal with JSON data without compromising scalability, availability, performance, full ACID support, and complete SQL functionality. With cloud-native development all around, JSON is becoming a more and more popular choice to store data, as it can be easily consumed by several programming languages and provides a persistent format for application objects – another reason being that JSON is schema-flexible, so applications can change over time to accommodate new types of data without having to modify backend data definitions. This lets you quickly react to changing application requirements without requiring you to normalize data into relational tables and with no restriction to changing data structure or organization at any time.

With AJD, your JSON document-centric applications typically use **Simple Oracle Document Access (SODA)**. SODA is a set of NoSQL-style APIs that help create and store collections of documents in JSON format and eliminate the need for SQL expertise for retrieving and querying JSON data. SODA collection APIs are exposed in several forms:

- Database tools, SQL Developer Web, and SQLcl

- SODA REST services

- Programming language drivers for Java, Node.js, Python, C, and PL/SQL

Quick note

AJD is similar to ATP with largely equivalent functionality and the same performance characteristics.

AJD provides all of the same features as ATP but with important limitations as *you can only store up to 20 GB of data other than JSON document collections*. There is no storage limit for JSON collections though. This could be a possible reason why AJD is offered at a lower price than ATP.

You can promote an AJD service to an ATP service to remove the 20 GB restriction on non-JSON data. You can not convert AJD to ATP, however.

AJD uses document-based databases and provides most of the same benefits that are typically associated with NoSQL document stores:

- **High availability and performance at scale**: Transparently scale the compute and storage capacity of your database while maintaining millisecond latencies for reads and writes.

- **Simple document APIs**: The SODA APIs make it easy to store JSON natively in the database. Using these APIs, you can build an entire application without having to write SQLs.

  As I said, AJD is much more than a simple document store. It provides a rich set of features that are typically not found in NoSQL databases.

- **Automatic administration and performance tuning**: Routine database administration tasks, such as provisioning, performance tuning, encryption, patching, and taking backups, are performed automatically, so you can focus on developing your application.

- **Full SQL query support**: Natively-stored JSON using the document store APIs is fully accessible using ISO standard SQL. With AJD, you can use SQL to perform real-time analytics over JSON collections. You can also create real-time relational views over JSON to expose collections to existing relational tools and applications.

- **ACID transactions**: AJD supports robust transactions over JSON collections. ADB follows ACID protocols for JSON datatypes, like other RDBMS workloads, and makes it easy to perform complex operations over multiple collections atomically.

- **Advanced security**: Encryption and data-safe options are available with AJD.

- **Supporting tools and services**: AJD comes with a number of supporting services for processing and accessing data, including the following:

  - **Oracle REST Data Services (ORDS)**: This can be used to build custom REST services over your JSON data.

  - **APEX**: This is used for building low-code applications over JSON collections.

  - **Oracle Machine Learning Notebooks**: This enables data scientists and data analysts to explore JSON data visually and develop analytical methodologies.

Keep one thing in mind: any SODA collection within AJD will have only JSON data. It cannot be mixed with LOB documents, unlike ATP databases.

Autonomous Data Guard is available for AJD. A standby database can be enabled either in a local region, cross-region, or both based on availability requirements. With Autonomous Data Guard, both the primary and standby (local or remote standby) databases are monitored for transactions and take the following actions:

- In case of the failure of the primary database, the standby database is converted into the primary database without user invention and with minimal interruption. Once failover is completed, a new standby database is automatically created by Autonomous Data Guard.

- Additionally, application or database admins can perform a manual switchover operation to convert the primary database into a standby database and vice versa.

Note that this is for shared infrastructure ADB only.

Now, since we have already looked at the classifications of ADB based on workload types, we will explore which infrastructure option is appropriate for your workload in terms of deployment.

ADB infrastructure deployment choices – shared and dedicated

Autonomous databases can be deployed on either the shared or dedicated Exadata infrastructure available in OCI. Shared Exadata infrastructure provides the smallest minimum commitment to get started, while dedicated Exadata infrastructure provides customers with more control over infrastructure operations, such as controlling the software version and update schedules.

There are several considerations for deciding which one to pick. We will go through them later in the chapter. For now, we will try to understand these services' classifications on a high level.

Shared

The shared deployment offering is a fully managed multi-tenant environment on a shared infrastructure, powered by Exadata-engineered system hardware on the backend. This environment is shared with several customers but Oracle makes sure you get the needed isolation in terms of resources to run your workloads, such as CPUs and storage. The provisioning is super simple to deploy with very minimal input and customers can get started immediately upon provisioning. It's also an effort to standardize the database environment for customers, including patching, without worrying about multiple life cycle environments, such as database software, versioning, environment separation, and so on. You can see that this environment is very simple to use, is commercially attractive, has no major commitment, has a flexible pricing model, which uses per-second billing; and can be configured for automatic scaling on demand.

Shared infrastructure is great for customers who want to be database users without worrying about any database operations, including software updates. Oracle has tried to bring uniformity, simplicity, and standardization, which can be considered the key to extreme scalability and performance. It is an ideal choice for a line of business, departmental applications, or data marts, as well as making an excellent environment for developers or data scientists. You can start to deploy your workload in a shared offering of ADB and transition to a dedicated Autonomous Database when you are done with all sorts of migration, testing, and user acceptance.

> **Quick note**
> For ADB on shared infrastructure, the minimum database version is 19c.

Dedicated

With dedicated infrastructure, customers have their own dedicated Exadata infrastructure in OCI. It is a little more customizable than the shared infrastructure offering and intentionally created by Oracle to cater to those customer bases that require more control for IT, DBAs, and storage admin groups within an organization. The capacity of dedicated infrastructure is controlled by customers for their workload. This offering provides completely isolated environments to customers, which could be a requirement for them. It provides an opportunity to create one or more container databases hosting thousands of PDBs within it.

The overall idea behind the dedicated deployment model is its customizable policy for the databases. The question comes – why does Oracle allow customization to this environment? To understand the answer to this, consider customer bases that are using Oracle databases for their workloads. There are customers with thousands of databases and several life cycle environments to manage. Not only that but certain applications could also be sensitive to database versions. Customers also want to control the update frequency, timing, and segregation of databases based on their behavior and workload types, which demands control of the database infrastructure. Consolidation is another important factor that is considered in dedicated offerings. These customizations provide flexibility to customers – yet Oracle manages operations that are repetitive in nature, such as backup, provisioning, patching, and so on.

> **Quick note**
>
> For autonomous databases on dedicated infrastructure, users have the ability to choose to patch Exadata infrastructure or container databases immediately via the **Patch Now** option. They can also reschedule an already scheduled Exadata infrastructure maintenance via the **Edit** options.

The following screenshot shows the flexibility of controlling the maintenance schedule in a dedicated infrastructure deployment. Customers can control this behavior based on their operational requirements.

Figure 1.4 – Maintenance schedule option for ADB

A dedicated ADB offering is available for both ADW and ATP workloads. A customer can have a dedicated ADB deployment and have either or both of these workloads running within it based on requirements. You can consider this as your own infrastructure running in the Oracle public cloud with services that are most concerning automated, such as provisioning, performance tuning, scaling, patching, database encryption, and backup and restore. As you know, Oracle Database has several options, such as RAC, Database In-Memory, Partitioning, Advanced Compression, and so on, to name a few, which are available within ADB offerings. Apart from this, additional Enterprise Manager Packs, such as Diagnostics Pack, Tuning Pack, RAC, and so on, are available within an ADB deployment.

There are broadly three logical roles that are involved in setting up a dedicated Autonomous Database: fleet administrators, DBAs, and database users. Depending on the separation of roles required for the deployment, you can have a specific role assigned or be part of multiple roles. Fleet administrators are mostly related to management aspects, such as provisioning the infrastructure, managing container resources within monitoring, and managing VM clusters within ADB dedicated on Exadata. A DBA, on the other hand, is responsible for creating, monitoring, and managing Autonomous Databases. These administrators have the ability to create application users, do backups, and restore configurations. The third type of user, database users, are mostly developers who are focused on application development and concerned about making a connection to the database for their work and queries. These users need not be part of Cloud Console users, as they can get connectivity information and required security wallets from an administrator in order to make database connections.

> **Quick note**
>
> ADB dedicated provides 99.995% SLA, which translates to <2.5 minutes of downtime per month. It is fully isolated from other tenants, as the VCN is hardware-enforced.

ADB on Exadata Cloud@Customer

As we saw with a dedicated offering for ADB running in the Oracle public cloud, this Cloud@Customer offering is different in the sense that ADB runs at a customer data center, behind their firewall. It also uses the same Exadata Database Machine infrastructure with networking configuration allowing Oracle to manage from the Oracle Cloud control plane. This option was brought by Oracle for customers who have strict data sovereignty requirements.

The most common reasons for using the **Exadata Cloud@Customer** (**ExaC@C**) model include the following:

- Corporate policies and regulations
- Network latency
- Integration needs
- Compliance and risk

Sometimes, it is hard for companies to move their workload to the public cloud, mostly because of country laws related to handling data. System integration needs could be more complex where several applications, hardware, and so on need to be tightly integrated. Network latency requirements also play a major role. Adopting a hybrid cloud strategy is another consideration, which plays a major role when considering cloud adoption. By keeping these in mind, Oracle came up with this offering under ADB on its engineered system platform for customers. Certain key facts about dedicated ADB deployments are listed here:

- The database offering named Oracle ExaC@C is not very new and was already announced in 2017. This option provided cloud flexibility to customers who were not able to move to the public cloud.
- With an ADB offering, Oracle was able to bring cloud self-service and a pay-per-use financial model to customers.

- Customers get the full advantage of ADB along with the benefits of dedicated infrastructure in their own data center.

- You can run a mix of **Autonomous Transaction Processing-Dedicated** (**ATP-D**) and **Autonomous Data Warehouse-Dedicated** (**ADW-D**) databases on the same dedicated ADB Exadata rack.

- Services can be managed via the OCI Console, the CLI, and APIs.

> **Quick note**
>
> There are a few key differences around networking configuration and backup support when you compare dedicated ADBs in the public cloud versus Cloud@Customer. On OCI, networking is configured via a VCN by the customer, whereas, on ADB ExaC@C, it requires connectivity to the cloud control plane and the Exadata system primary and client networks to be set up, which the Oracle field engineer will validate on-site.

- **Backup locations**: On OCI, only object storage is supported. On ExaC@C, customers have the option to back up to object storage, local Exadata storage, their own network-attached storage (NFS), or **Zero Data Loss Recovery Appliance** (**ZDLRA**), also known as **Recovery Appliance** (**RA**).

> **Quick note**
>
> ZDLRA is Oracle's optimized solution for database backup and recovery. This RA changes the way backups and recovery are carried out in any traditional database deployment. It enables incremental forever backups and efficient any **Point-in-Time Restore** (**PITR**). The purpose of this appliance is to enable zero data loss functionality for RA backup destinations. The customer has the ability to choose to enable Real-Time Redo Transport in RA-configured environments, which enables the database to automatically ship all redo logs in real time to RA for a zero data loss (less than 1 second) recovery setup.

I believe that now that we have looked at every classification of ADB in terms of infrastructure type, you can make a decision based on your workload requirements to either go with shared or dedicated options. Let's try to understand some of the merits to consider while considering this move with ADB.

Understanding why to use an Autonomous Database

There has been a major change starting with Oracle Database 9i, where Oracle started to introduce and enhance many automation capabilities, from memory management to workload monitoring and performance tuning. All of these set a solid foundation for autonomous capabilities and are used in ADB. Not only database automation but also database infrastructure with engineered systems was invested in and brought in by Oracle, which is considered the best platform for the Oracle Database workload, as these systems are preconfigured, pretested, and optimized converged platforms for the database. Take a look at the following automation capabilities over the span of a decade to get a view of the enhancements in Oracle Database:

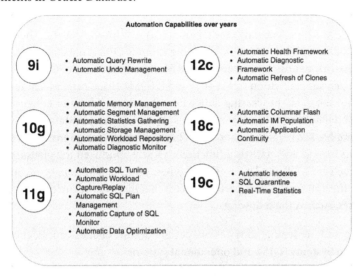

Figure 1.5 – Depicting Oracle Database automation over the years

In addition to these innovations, Oracle has spent more than 10 years automating the database infrastructure:

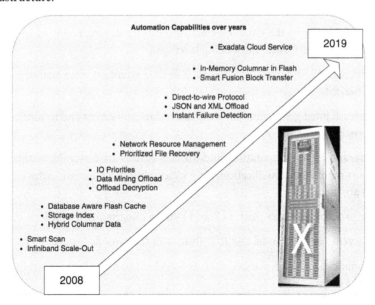

Figure 1.6 – Depicting Oracle's infrastructure automation over the years

Decades of innovation in the ML capabilities of Oracle Database

After all these innovations, let's see how the Oracle database uses ML and AI right inside the database.

In Dec 2019, Oracle Machine Learning (earlier known as Advanced Analytics) became a *free* feature of every Oracle database. Oracle Machine Learning extends Oracle database(s) by delivering more than 30 in-database ML algorithms and automated ML functionality via SQL APIs (OML4SQL) and integration with open source Python (OML4Py) and R (OML4R). Oracle Machine Learning "moves the algorithms, not the data," processing data where it resides. It further enhances the benefits of Oracle's multi-model converged architecture by supporting various data types and data models such as Spatial, graphs, XML, JSON, and so on. Through innovation in infrastructure software and various algorithms for ML and statistical functions, Oracle optimized the platform to run different workload types, such as OLTP and DSS workload types. Without additional cost, customers can take full advantage of the enterprise-class performance, reliability, and security of Oracle Database for particular use cases, such as the following:

- Processing and analyzing all types of spatial data in business applications, **Geographic Information Systems (GIS)**, and operational systems

- Using graph analysis to discover relationships in social networks, detect fraud, and make informed recommendations

- Building and deploying ML models for predictive analytics

Let's take a look at various capabilities related to the uses of ML in ADB:

- ML is integrated across the database stack and infrastructure, resulting in the autonomous capabilities of the full Oracle stack being leveraged.

- By introducing ML, different personas such as data scientists, data analysts, developers, and DBAs/IT benefit.

- Complete stack intelligence eliminates expensive data movement and minimizes access latency between systems.

- Data exploration, data preparation, and ML algorithms are faster. More than 30 algorithms that support regression, classification, time series, clustering, feature extraction, and anomaly detection are included.

- ADB supports open sources, such as R and Python integration, useful for a data scientist.

- You can develop an ML model and R/Python with the provided tools.

- ADB automates key ML processes, such as workload patterns, auto-indexing behavior, SQL profiles, and so on.

- ADB supports existing enterprise backup, recovery, and security mechanisms and protocols.

- Most importantly, it brings the algorithms to the data.

With Oracle Machine Learning, along with Oracle Database and infrastructure capabilities, Oracle has evolved as a fully converged data management platform.

For any modern organization, databases play a critical role in storing important application and business information and are essential for the efficient operation of an organization. Databases are managed by DBAs and they are often occupied with operational things such as backups, patching, and performance tuning, and spend the majority of their time on this kind of manual task to make sure the database is up and running. Any DBA errors because of manual work can lead to a serious impact on database availability, performance, and security.

On the other hand, today, several data architecture challenges exist – several customers try to experiment with many different kinds of databases, such as relational, document store, graph and spatial, time series, NoSQL, and Big Data analytics to name a few. It brings several administration challenges:

- Hard to enforce consistency of data across applications
- Hard to do consistent backups for different databases
- Issues with cross-database compatibility
- Separate security patches for every database
- Separate upgrade cycles for databases
- Varied level of performance levels across databases
- Varied availability – different ways to implement disaster recovery
- Complex integration for analytics
- Hard to match varied skills for different databases

These challenges can be easily addressed with ADB because of its **unified, simplified, and standardized approach.**

If we explore it further, we see that failing to apply a patch or security update can leave open vulnerabilities in databases. With massive data growth, it brings more complexity, making them even more difficult for DBAs to manage, secure, and tune for maximum performance. Needless to say, databases that are not performant or slow-running can negatively impact employee productivity and impact customers. Poorly designed disaster recovery plans can lead to huge business impacts.

ADB uses ML in three key areas of ADB: diagnostics, recovery, and optimizations for each layer of the deployment stack, each respectively corresponding to the following:

- Inside database infrastructure, ADB automatically detects and recovers from failed servers, storage, or network switches or links

- Automatically detecting an anomaly or hangs in the database behavior, comparing these behaviors with known issues, and automatically identifying known issues and fixing them or automatically opening a **Service Request (SR)**

- Automatically optimizing customer workloads with real-time statistics and automatic indexing

So far, we have explored the generic benefits of using ADB.

Because ADB runs on Exadata systems, RACs are also provisioned in the background to support the on-demand CPU scalability of the service. This is transparent to the user and administrator of the service. For all ADB offerings, there is an option of creating an optional remote standby database (Autonomous Data Guard) for automatic failover capabilities and disaster recovery purposes.

As mentioned, ADB runs on Exadata systems, but no Exadata installation, configuration, or management needs to be done or can be done. The `init.ora` parameters (database initialization parameters) are configured automatically in an Autonomous Database depending on the service selected – ADW, ATP, or AJD. Memory, parallelism, concurrency, number of sessions, and other parameters are automatically configured based on the number of CPUs allocated to the service. Most of these parameters cannot be modified, and the few that can be modified should only be done for very specific reasons by qualified DBAs. This is discouraged in ADB, as it defeats its purpose.

> **Quick note**
>
> When an Oracle instance starts, the very first file it reads is initialization parameters (`init.ora`) from an initialization parameter file. As of today, customers do not have the capability to set database parameters or resource manager configs while creating ADB instances or modifying them via the Console, APIs, or an SDK after the instance is created. All parameters are set to optimal values based on the specific workload type and this may be different from regular database defaults. The ATP and ADW have two different resource management plans. The ADW plan name is `DWCS_PLAN` and the ATP plan name is `OLTP_PLAN`.

Tablespace management is performed automatically by ADB and cannot be changed by the customer. Customers have full access to view the information of the space allocated to their instance, but it cannot be changed. The only input the customer needs to provide is the number of terabytes of data they would like the database to be able to hold. This number can be increased or decreased in real time. ADB handles adjusting the data location based on this user setting. The default tablespace is the same for ATP and ADW.

Loading and maintaining data in ADB can be done as one-time loads, best when staged through Oracle Object Store, or as continuous data ingestion or synchronization with other sources. ADB supports three object stores and can read and write directly to these three. The supported object stores are Oracle Object Store, Amazon S3, and Azure Object Store. Object stores are ideal for staging export dump files that are going to be imported into the Autonomous Database.

The same applies to flat files that would be loaded into the database. An ADB supports the Oracle Database external tables feature – so flat files on the object store can act as autonomous external tables. Please note that it is best to host these tables on Oracle Object Stores that are fast and connected to the Autonomous Database to reduce latency and other issues around access time to database objects.

Also available for transactions and data work location in real time or near real time, or to maintain synchronized copies of databases, is Oracle GoldenGate, which can be configured with an Autonomous Database as a target database. This allows ADB to become a full replica copy of another database for uses such as reporting, disaster recovery, development, testing, and QA. Once a decision is made to move the database to ADB services, the next step is to determine what to do with the application accessing that database.

Just like rehosting the database in the cloud, rehosting the application to the cloud may have its own benefits. If the application using the Autonomous Database is an existing application, there are two preferred options for hosting the application. The first option is to keep the application in its existing environment and replace the existing database with access to the Autonomous Database. The second option is to rehost the application to OCI. Rehosting the application may be straightforward or may require substantial reconfiguration.

Autonomous Database value proposition

OCI provides the required elasticity, agility, security, and global reach, and helps customers focus on their workloads and not the infrastructure. With OCI DbaaS, OCI is helping customers lift and shift their on-prem database workloads to the cloud. While OCI continues to add more tooling to the OCI DBaaS, customers want complete automation and no management overhead. They want their database service to be fully managed, where routine tasks such as patching are automatically done, backups are available to customers when they need it, systems scale automatically to workload requirements, and much more. ADB services such as ATP, ADW, and AJD aim to fully manage customer databases based on customer preferences. All exceptions and failure cases are automatically managed by Oracle.

The key principles of the ADB service are as follows:

- **Self-driving**: Customers define the service level and ADB makes it happen
- **Self-securing**: The service protects against external attacks and malicious internal access
- **Self-repairing**: The service enables higher availability with automated protection from downtime

The key customer benefits of ADB services are as follows:

- Reduced services costs and increase productivity for customers.
- Reduced admin costs, with complete automation of operations and tuning.
- Reduced runtime costs by dynamically adjusting resources and eliminating underutilization.

- Deploying new apps in minutes, with faster **Time to Install** (**TTI**) and faster **Time to Deliver** (**TTD**).

- Reduces the cost of downtime – less than 2.5 minutes per month.

- Elastically grows and shrinks compute or storage without downtime. Pay only for what you use.

- Eliminates human labor to provision, secure, monitor, backup, recover, troubleshoot, and tune.

- Automatically upgrades and patches itself while running. Testing automation ensures changes are safe.

- Protection from external attacks and from malicious internal users.

- Protection from attacks by automatically applying security updates with no downtime.

- Automatic encryption of all data.

- Most reliable.

- No human labor, and hence no human error

- Zero downtime with scale on-demand.

- SLA 99.995% guarantee available with the **Active Data Guard** (**ADG**)- **Dedicated** option. SLA 99.95% guarantee available with the **ADG-Shared** option.

- Automation eliminates administrator errors.

- SLA guarantees 99.995% availability. < 2.5 minutes of downtime per month, including planned maintenance.

So far, we have seen how Oracle Database has evolved over time to bring automation capabilities along with innovation in infrastructure and this innovation is continuing with each release. The current release of the Exadata platform, X9M, is the fastest database machine engineered to run ADB. Now, let's review a few use cases and considerations for ADB.

Reviewing use cases for ADB

Customers who want a highly performant database service capable of automatically managing life cycle operations would use ADB. Things to remember – ADB provides **no host access and no OS customization capabilities**. Customers who are looking for custom requirements and host access will use OCI DbaaS services, where customers can choose from VMs (DbaaS VM images), BM options (DbaaS bare metal services), and Exadata shapes, which include **Exadata Cloud Infrastructure** (**ExaCS**) and ExaC@C deployments within the Oracle Cloud Infrastructure portfolio of services.

Let's discuss having no access to the host OS and no OS customization capabilities a bit. This means a lot when it comes to operational efficiency and standardization. No root or sysdba logins are allowed – the only logins allowed are admin, privileged default ADB user, or regular database user. It means no call-outs to the OS – thus, preventing installation or modification of any software on the system. It is sealed from external administrative access and powered by AI. ADB eliminates any possibility of misconfiguration, database vulnerabilities, malicious activities, and human errors. Database clients can connect securely using a TLS wallet. Oracle manages all the operational aspects to make sure you focus on your application and leave SLAs, patching the OS, database upgrades, security, and several performance-tuning aspects to be part of the autonomous capabilities of the offering.

Some of the stats on database manageability are as follows:

- More than 39% of DBAs handle 50 or more databases (source: *From Database Clouds to Big Data: 2013 IOUG Survey on Database Manageability*)

- Challenges with the typical patch management process – it's complex, time-consuming, and involves dependency on multiple stakeholders

- Most companies face high downtime because of a lack of standardization across database fleets within an organization

- Building high availability for databases takes significant time, effort, and expertise, and an enterprise needs it at the click of a button

---

Quick note

The **Independent Oracle Users Group** (**IOUG**): The major focus of the IOUG is on Oracle technology and database advocates. It promotes empowerment through education so that the users can be more productive in their work related to these technologies and also help take better business decisions through providing technology direction and networking opportunities, sharing best practices, and delivering education.

Now, let's go over some typical considerations that a customer can evaluate when implementing an Autonomous Database. Unlike on-premises deployments, many steps are not needed with Autonomous Databases and because of this, it is easy to deploy as well. Still, there could be several considerations, such as the level of automation and functionality required, workload characteristics of databases such as ATP, ADW, or AJD, provisioning and loading data to Autonomous Databases, connecting your applications to them, and many more.

Oracle ATP supports all operational business systems, including both departmental as well as mission-critical applications, but unlike other cloud providers, ATP doesn't just support one transaction processing use case; it can also support mixed workloads where you have a mixture of transaction processing, reporting, and batch processing, making it the perfect platform for real-time analytics based on operational databases. This enables users to get immediate answers to any question.

Integrated ML algorithms make it the perfect platform for applications with real-time predictive capabilities. Advanced SQL and PL/SQL support make it the perfect platform for application developers, as developers can instantly create and effortlessly use ATP, eliminating any dependence and delays on others for hardware and software. The fact that it's self-tuning also eliminates any need for database tuning and accelerates developer productivity. With the availability of APEX within ATP, you can deploy applications faster for their use case or modernize legacy applications with them.

Oracle ADW supports all types of analytical warehouses and decision support database workloads. ADW is particularly well-suited to creating new dependent or independent data marts that allow analytical projects to be started easily. It is a good environment for sandbox experimentation on the part of data scientists, sifting through data and storing large amounts of data and data lakes. It includes analytics and visualization tools, Oracle Machine Learning and Oracle Data Visualization Desktop, and provides an end-to-end environment for application development, data analysis, and fast, flexible database services.

Understanding the business benefits of using ADB

Now that you have an understanding of Autonomous Databases and are considering migrating your database workload to the ADB platform, the very next thing that comes to your mind is what the business value is – potential savings, TCO, ROI, and more. Well, all these are valid concerns, as technical merits are not the only decision-making factor when considering the cloud as a platform for workload migration. You might be interested in comparing the cost of business as usual versus operating from the Oracle cloud. You might need to create a TCO/ROI analysis before a decision can be taken by the LOB or senior management. Oracle's Data Management Cloud Services team provides a tool to provide guidance in this regard. A TCO calculator was made publicly available by the Oracle team to help organizations or individuals see a detailed breakdown of the potential cost savings that they would realize if they moved their database workload deployments to Oracle's Database Cloud Service. The TCO calculator provides a side-by-side comparison to an equivalent on-premises deployment or competitive deployments, as well as a detailed report of potential cost savings in terms of compute, storage, software, and facility expenditures. We will look at the steps to calculate a TCO once we understand the various metrics that impact the TCO.

We can look at several metrics that can help you understand TCO comparisons with any traditional deployment or other cloud vendors:

- Improved business performance gains – operating margin improvement due to targeted end process (business) improvement(s)
- Improved DBA and system administrator productivity
- Reduced development cycle time
- Reduced IT infrastructure acquisition
- Reduced IT infrastructure maintenance and support

- Reduced non-compliance risks – data

- Reduced planned and unplanned downtime

- Reduced security and data breaches

We will discuss some of these benefits in detail in the next sections.

Improved business performance gains

Autonomous Databases can help drive business improvement efforts:

- Customers focus on their data and application without worrying about any configuration in terms of infrastructure or database software such as RDBMS and Grid Infrastructure software. ADB automatically handles creating the ADW, ATP, or AJD offering (for example, the compute, storage, and network configuration), securing data (for example, encryption by default at rest and in transit), backing up the database, patching and upgrading the database without downtime, and scaling the database up or down (without impacting system performance or customer experience). As a result, customers can quickly deploy the database while freeing up technical staff to assist with business insight (rather than maintaining infrastructure).

- Second, Autonomous Databases can easily integrate into a variety of on-premises, cloud, and hybrid systems, allowing firms to securely consolidate information for analysis and reporting. Additionally, ATP also supports both relational and non-relational data models to reduce data fragmentation and management issues resulting from siloed data stores.

- Finally, on-demand scalability lowers the threshold for starting and growing data warehouse projects, so proof-of-concept and pilot projects can occur without major financial investments and related approvals. ATP provides in-database ML algorithms to make it easier to build ML models and analytical dashboards without moving data out of the database, resulting in improvided business insight.

Improved DBA and system administrator productivity

By design, Autonomous Databases can increase DBA productivity as follows:

- With ADB, customers do not need to configure or manage any hardware or install any software. Database creation, backups, patching, security/encryption, and scaling up or down all are automatic. For example, all OS and SYSDBA operations are carried out automatically, settings are tuned, and errors are diagnosed and remediated. By continually monitoring the cluster for unusual changes, Oracle can compare any anomalies against known issues and automatically apply a fix (or create a service request and escalate as appropriate).

- ADB provides automatic downtime protection with high availability built into every component and proven Oracle database technologies (for example, RAC, Exadata, and so on) and HA features such as Autonomous Data Guard and cloning.

- ADB does not require manual tuning. Whether it is ADW or ATP, customers just need to load the data using a traditional method, such as Data Pump or GoldenGate, and they can get going with application queries without worrying about specific partitioning schemes, parallelism, indexing, compression algorithms, and so on. Based on the workload behavior (ADW or ATP), the database is automatically configured for high performance, leveraging ML at the backend. ADB continuously optimizes memory, data formats, indexes, parallelism, queries, and so on for each workload using ML.

- ADB offers dedicated cloud-ready migration tools for easy migration from Amazon Redshift, SQL Server, and other databases. As with other Oracle cloud solutions, ADB is fully compatible with existing Oracle on-premises data management workloads, so customers can leverage existing infrastructure and knowledge.

- Best practices for performance, availability, and security can be consistently and automatically implemented from the beginning.

Reduced development cycle time

Autonomous Databases help reduce development time:

- Architecture is the same for both the cloud and on-premises; there is also support for familiar development and administrative tools, such as Oracle SQL Developer, Data Pump, and SQL*Loader, so developers don't need to learn about new ones when they use ADB.

- Customers get full interoperability with their on-premises Oracle databases, as well as integration with other Oracle (for example, Analytics Cloud) and third-party cloud services (for example, Amazon, Azure, various BI/analytics tools, and so on); migration tools are also provided for all major database providers. ADW supports multi-model data and workloads (e.g., analytical SQL, ML, graph, and spatial). ATP supports both relational and non-relational data models, along with Oracle's low-code application development platform (APEX).

- Autonomous Databases do not require any tuning. ADB is designed as a "load and go" service: you start the service, define tables, load data, and then run queries. Customers do not need to consider any details about parallelism, partitioning, indexing, or compression. The service automatically configures the database for high-performance queries with ML optimizations.

- Customers can work directly with their data by connecting via any of the available Oracle client language libraries, including Oracle Net (SQL*Net), JDBC, and other drivers.

- The ability for developers and related groups to autonomously provision and elastically scale databases, rather than waiting weeks or months for traditional infrastructure acquisition, installation, and provisioning, can significantly improve productivity.

Reduced non-compliance risks – data

Autonomous databases provide enterprise-class security:

- An autonomous service, so it always runs the latest security patches and avoids the inevitable delays when administrators try to maintain hundreds or thousands of servers.

- Data is automatically encrypted by default in the cloud, as well as in transit and at rest, with customer-controlled keys. Data at rest is encrypted by default using TDE, and only authenticated users and applications can access the data when they connect to the database. Connections to ADB are made via SQL*Net. TCP and TCP with SSL security protocols are supported methods. TCP with SSL uses certificate-based authentication and the SSL security protocol. Using this mechanism, ADB ensures secure communication between the Oracle database and any clients that are connected to it and prevents any breach of the confidentiality of data during communication because of the encryption in place.

- ADB uses strong password complexity rules for all users based on Oracle Cloud security standards.

- Oracle's autonomous cloud services offer adaptive intelligence-enabled cyber threat detection and remediation.

- Administrative access privileges are greatly reduced, increasing resiliency against human errors, malicious insiders, and hackers.

- Security best practices can be consistently implemented from the beginning. By improving the overall security and offering the agility to meet new requirements, ADB can help reduce non-compliance security risks.

Reduced planned downtime

Autonomous Databases offer excellent availability features with a 99.995% SLA (less than 2.5 minutes of unplanned and planned downtime per month).

ADB is based on proven, enterprise-class Oracle technologies and solutions (for example, RAC, Exadata, etc.), offering high availability characteristics and redundancies. Secondly, patching, updates, and backups are automatically applied without service interruptions; the same is true for the on-demand scaling up/down of compute and storage resources. In fact, since all aspects of the service are autonomously handled on a standardized, extremely scalable platform operated by Oracle, planned maintenance activities can be more easily handled, and best practices (for example, rolling patching) are consistently applied from the beginning.

Reduced unplanned downtime

Autonomous Data Warehouse Cloud Service (**ADWCS**) offers excellent availability features with a 99.95% SLA. First, ADB is based on proven, enterprise-class Oracle technologies and solutions (e.g., RAC, Exadata, Autonomous Data Guard, etc.) offering high availability characteristics and redundancies. Secondly, since ADB is fully compatible with on-premises Oracle databases and all existing applications, it is less likely to experience unplanned downtime.

Thirdly, since all aspects of the service are autonomously handled on a standardized platform operated by Oracle, the chance for human errors, as well as actions by malicious insiders and hackers, is greatly reduced. Moreover, availability best practices are consistently applied from the beginning. Equally, patching, updates, and backups are automatically applied without service interruptions; the same is true for the on-demand scaling up/down of compute and storage resources. Finally, ADWCS offers a single contact point for end-to-end support, which can further reduce the amount of time spent troubleshooting issues.

Reduced IT infrastructure acquisition

Autonomous Databases provide virtually unlimited scaling capacity with no upfront hardware and software costs for the hardware involved with either transaction processing, analytics, JSON, or mixed workload database infrastructure. Other than paying for an initial subscription, there are no other upfront costs. Additional resources can be purchased as needed in any combination of computing and storage sizes (that is, no rigid shapes). ADB offers both serverless and dedicated deployment options on ECS, so customers can balance costs and other considerations (for example, performance, security isolation, software control, and so on). ADB offers instant scaling up/down of compute or storage independently of each other with no downtime. This elasticity avoids the need to acquire and provision transaction processing and mixed workload infrastructure potentially months or even years in advance, thereby shifting customers to a cost model driven by actual usage, rather than longer-term projections.

Further, ADB offers proven Oracle features and solutions including Active (or Autonomous) Data Guard, RAC, multitenant options, and Exadata infrastructure. With these capabilities, customers can conserve bandwidth, improve performance, and reduce cloud requirements to further decrease costs. Finally, since Oracle architectures, standards, and products are similar for the cloud and on-premises, migrations, integrations, and other changes are relatively seamless, allowing transparent movement between the two. As a result, a firm's existing transaction processing and mixed workload infrastructure can be reduced and simplified.

Reduced IT infrastructure maintenance and support

ADB provides a virtually unlimited scaling capacity without many of the ongoing infrastructure maintenance and support costs for traditional on-premises or third-party hosting environments involved with transaction processing and mixed workload database infrastructure. Firstly, with its elastic capacity, ADB customers avoid maintaining and supporting the under-utilized infrastructure that typically exists at their own sites. The compute and storage capacity can be adjusted independently. Secondly, ADB operates at a scale that isn't achievable for most firms, providing further cost advantages. ADB offers both serverless and dedicated deployment options on ECS, so customers can balance costs and other considerations (e.g., performance, security isolation, software control, and so on). Thirdly, ADB offers proven Oracle features and solutions including Active (or Autonomous) Data Guard, RAC, multitenant options, and Exadata infrastructure. Finally, since Oracle architectures, standards, and products are similar for the cloud and on-premises, related knowledge and procedures are easily transferable. Together, these features and capabilities help reduce maintenance-related costs.

Steps to calculate the TCO

Please perform the following steps to calculate the TCO/ROI for your use case:

1. First, go to the TCO tool at the following link: `https://valuenavigator.oracle.com/benefitcalculator/faces/inputs?id=408D37F02AF0A012283B0D29A4A0708A`.

2. Once there, click on the **Start Now** arrow.

3. Now, you will need to input your company name, select a location, and a currency type.

4. Review and generate the report.

Let's review each step in detail. Open the following URL in a browser: `https://valuenavigator.oracle.com/benefitcalculator/faces/inputs?id=408D37F02AF0A012283B0D29A4A0708A`. This will open up the ADW TCO page as shown in the following screenshot:

Figure 1.7 – TCO tool page for ADW

When you click on **Review your figures**, it presents you with a TCO calculation based on certain assumptions and also provides an opportunity for you to edit assumptions for your specific scenarios. For example, you can choose either **Bring Your Own License** (**BYOL**) or **License Included** for TCO calculations. Many other parameters can be changed to suit your needs.

In the next screenshot, you can see a place to describe the environment, such as the number of cores available in computing, the size of storage, which can be calculated based on usable or raw storage capacity, and so on:

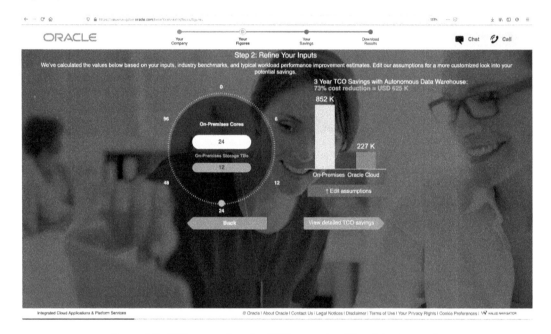

Figure 1.8 – TCO comparison for ADW versus traditional deployments

For example, *Figure 1.9* compares on-premises TCO versus the cloud TCO for ADB and shows a 70% cost reduction; if you look at the **TCO** section, it shows cost distribution across software, compute, storage, network, and labor. The **Environment** section compares cores or OCPUs and storage across two environments.

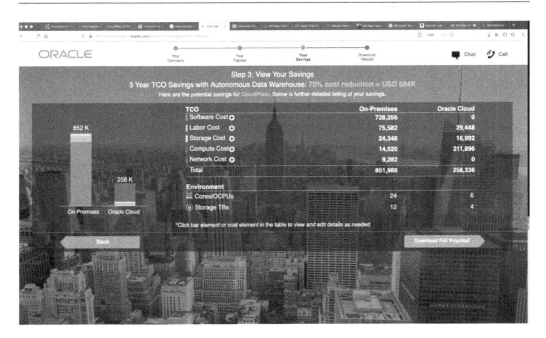

Figure 1.9 – TCO savings with ADB

The following screenshot shows the option to download the report for TCO for your use case. Inside the downloaded report, you will see input used for cost calculation and cost element per line item.

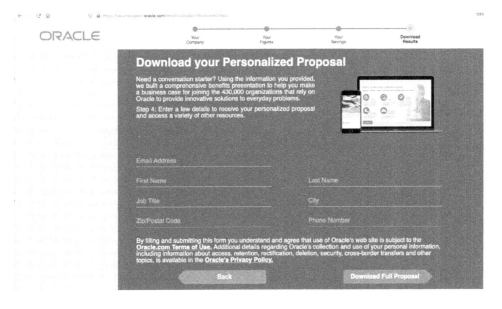

Figure 1.10 – TCO proposal download for ADB

It also summarizes cost calculation across storage hardware, software, facilities, and productivity. This is shown in the following screenshot as an excerpt from the report:

Figure 1.11 – Three-year TCO savings representation with ADB

From this screenshot, a comparison of business-as-usual deployment versus that of Oracle Cloud clearly shows two things: reduction in cost and fewer resources needed on autonomous services compared to a traditional deployment because of Exadata-based database capabilities. It is a great saving for customers and helps them make the decision to move their workload to the cloud with data points in hand.

With the TCO and ROI calculator report, the next step would be to get costing involved based on the **Stock Keeping Unit** (**SKU**) of the Autonomous Database and associated storage, and so on. Let's see how you can easily create a **Bill of Material** (**BOM**) for your solution.

BOM and SKUs for autonomous databases

There are two components to Autonomous Databases:

- **Infrastructure component**: You need to subscribe to Exadata Cloud Infrastructure for an OCI deployment or ExaC@C infrastructure for a C@C deployment.

- **Database OCPUs**: Subscribe hourly per active OCPU consumed, using either **License Included** or the **BYOL** type. Billing happens monthly at an aggregate level, a total of all active OCPUs, for each Exadata Cloud Infrastructure resource.

There are two scenarios possible in terms of licensing: existing Oracle customers may have already invested in Oracle databases and options. Oracle provides flexibility to use BYOL while provisioning an Autonomous Database in the Oracle cloud. This option brings down the service cost significantly. The other option is the **License Included** service deployment where customers do not own prior licenses of Oracle databases.

> **Note**
>
> If you are using BYOL, it is feasible to consistently change from 16 or fewer OCPUs to more than 16 OCPUs. The main prerequisite is that your BYOL should incorporate RAC when scaling past 16 OCPUs. For the most accurate policies and the latest information, you should check Oracle's site.

ADB-D on OCI pricing details can be found on the ADB pricing pages. ADB on ExaC@C pricing details can be found on Oracle's website under the *Pricing* information page:

- `https://www.oracle.com/autonomous-database/`
- `https://www.oracle.com/autonomous-database/autonomous-transaction-processing/pricing/`
- `https://www.oracle.com/autonomous-database/autonomous-data-warehouse/pricing/`

On OCI, the license type is defined at the Exadata infrastructure level. On ExaC@C, this is defined at the Autonomous VM cluster level. As of today, you can only use one license type (**License Included** or **BYOL**) on an Exadata rack until multiple VM clusters per rack are supported on ADB-D and ADB-ExaC@C, respectively.

Summary

Well, I hope you have had a great start to your ADB journey toward making a decision to take your workload to OCI ADB. You learned about deciding which kind of ADB is appropriate for your use case. We looked at the differences between shared versus dedicated infrastructure. We explored use cases for both kinds of deployments. Getting a TCO and ROI calculation for business is important and we have seen how you can quickly create them for your business case. With the evolution of Autonomous Databases, IT leaders must think of modernizing enterprise computing by transitioning it to a cloud-based model to take advantage of lower costs.

ADB combines the flexibility of the cloud with the power of ML to deliver data management as a service. I hope by now that you have an understanding of what ADB is, its technical building blocks, the reasons why you should consider ADB, use cases for ADB based on deployment types, business benefits, and creating BOMs and the TCO/ROI for your business justification.

In the next chapter, we will cover various deployment choices in detail such as provisioning, IAM and networking requirements, and connectivity.

Questions

1. What are the available infrastructure options for ADB?

2. What is the difference between a shared infrastructure and dedicated deployment?

3. What are the methods to connect to ADB?

4. Are there any limitations of ADB?

5. What is backup retention for databases in ADB?

6. What are the provisioning options for ADB?

7. For BYOL, is it possible to seamlessly transition from 16 or fewer OCPUs to greater than 16 OCPUs?

8. What is the minimum requirement for OCPUs in ADB? Does it support per-second billing?

9. Can I use Enterprise Manager to monitor ADB Dedicated instances?

10. Does ADB support an XML feature?

11. What are the shapes of the Exadata Cloud Infrastructure racks that are available for ADB Dedicated on OCI?

12. What are the shapes of the Exadata Infrastructure racks that are supported on ADB on ExaC@C?

13. Can I associate a subscription to Block and Object Storage with ADB and not use Oracle ADB – Exadata Storage?

Further reading

- *Oracle Cloud Infrastructure for Solutions Architects* by *Prasenjit Sarkar*

Answers

1. The available infrastructure options are shared Exadata infrastructure and dedicated Exadata infrastructure.

2. A shared infrastructure deployment provides easy-to-provision Autonomous Databases where multiple tenants can share an Exadata Cloud infrastructure. With dedicated deployment, you can consider it your own reserved Exadata infrastructure in OCI, which provides more flexibility and control than shared options.

3. All standard connectivity methods, such as JDBC, SQL* Net, or any SQL client tools such as SQL Developer, are used for connectivity.

4. ADB is a fully managed service from Oracle. You don't have access to the OS; additionally, not all DBEE administrative features are available. You should look at the Oracle documentation, as features might change over time.

5. Up to 60 days (7, 15, 30, and 60 days).

6. OCI support both BYOL and a License Included model for provisioning.

7. Yes, you can seamlessly scale beyond 16 OCPUs, although one of the requirements enforced by Oracle is to have a RAC license for scaling beyond 16 OCPUs in a BYOL scenario.

8. You need at least one OCPU to provision ADB, although upon provisioning, you can shut down the instance and database OCPU billing will stop. Keep in mind that you will be responsible for storage as long as the service instance exists.

9. Yes, Oracle Enterprise Manager 13.3, along with EM DB plugin bundle patch 13.32.0.190731 can be used. Keep in mind that only ATP-D databases are supported by EM at moment.

10. Yes, ADB supports XML DB features with certain restrictions.

11. Dedicated shape options include a quarter rack, half rack, and full rack. Underlying hardware could be one of Exadata X7, X8, X8M, or X9M shapes.

12. Shape options include a base rack, quarter rack, half rack, and full rack. Underlying hardware could be one of Exadata X7, X8, X8M, or X9M shapes.

13. No.

2
Autonomous Database Deployment Options in OCI

In the previous chapter, you learned how to decide which **autonomous database (ADB)** is best suited for your workload. Now it's time to deploy your ADB in **Oracle Cloud Infrastructure (OCI)**. Before moving on to various deployment options for ADB, I'll introduce you to Oracle's Always Free Tier for OCI so that you can have a better understanding of the environment and visualize things in a better way. As we progress, you will learn about various deployment options available to users, such as serverless architecture (shared), the dedicated offering on ExaCS, and ExaCC platforms in OCI. More importantly, understanding the requirements of the network architecture such as the access checklist, private endpoints, VCN, subnetwork considerations, and considering IAM's work separation policies will ensure that your deployment is well architected, keeping in mind best practices and security.

In this chapter, we will cover the following topics:

- Prerequisites – OCI's Free tier account
- Deploying serverless – Always Free ADBs
- IAM, networking, and security for shared and dedicated deployments
- Deploying serverless shared – ADB (ADW, ATP, and AJD)
- Deploying dedicated ADBs (in OCI on Dedicated ExaCS, In ExaCC)

By the end of this chapter, you will know how to set patch schedules for your dedicated ADBs and the options available across these deployments.

Technical requirements

In this chapter, though there are no technical requirements, if you are familiar with OCI Console, you will be able to visualize the topics discussed. I strongly encourage you to subscribe to a Free tier OCI account and use the resources available in the cloud for free.

OCI's Free tier account

OCI's Free tier can be considered a promotional account that's provided so you can explore the concepts around IAM, networking, free ADB, compute, and storage resources for learning purposes. A Free tier account is available to individuals as long as they want to keep using it.

OCI Free Trial

OCI Free Trial is more flexible than the Free tier in terms of exploration and learning of cloud resources that are otherwise not available in promotional Free tier accounts. It provides $300 of cloud credit that has to be used within 30 days. This credit can be used against any eligible service available on OCI.

In order to sign up for a free trial, you need to visit the free cloud trial website (`https://www.oracle.com/cloud/free/`) and click **Start for Free**. The Oracle website shows you all the eligible services available to you as part of the trial.

> **Quick note**
>
> During sign-up, you should pay careful attention to the home region because this is where you can provision most of the Always Free resources, such as ADB.

Unlike most vendors, in order to validate users, OCI might ask you to provide credit card information and other information, such as a phone number. Oracle uses it for security and validation purposes. Please keep in mind that Oracle will not charge your credit card unless you decide to upgrade your account by explicitly logging into your cloud account and choosing to upgrade.

Once your 30-day trial ends, your account will still remain active, and you can continue to create and use the **Always Free** resources available to you. All the paid resources that you might have provisioned during your 30-day free trial will be reclaimed unless you choose to upgrade your account to a paid account. Free accounts remain available if you use the account and do activities such as resource usage within 60 days.

If you decide to upgrade to a paid account, you have the option of a **Pay as You Go** account with no commitment, or a monthly flex account that offers discounted pricing based on the commitment to usage.

Let's explore what is available to you as free resources in your promotional account. It will help you understand the availability of resources and the service limits to plan your learning on OCI.

Always Free resources on OCI

All resources that are available as always free to an individual or customer tenancy are labeled as **Always Free** and can be validated by logging in to your tenancy and checking into the Console. These resources are free of charge for the lifetime of the tenancy.

For a better understanding of what you can provision in OCI as a free resource, Oracle provides flexibility to create resources such as application servers, ADBs, storage to support compute VMs, object storage, load balancers, and networking components such as VCN, subnets, and security list. By using these resources, you can build an application as you see fit. These resources can help you learn the concepts, you can do things such as run a small application with a database in the backend or perform proof-of-concept.

Let's summarize the **Always Free** eligible resources available in your cloud tenancy that you might see in the Console:

- Compute VM (two VM/servers, AMD, or ARM processor)

- Databases: ADB (two free ADB instances)

- Load balancer for applications (one load balancer)

- Storage: Block volume (you have 200 GB of total block volume storage)

- Storage: Object (20 GiB for storing files, documents, logs, etc.)

- Vault (20 key versions of master encryption keys protected by a **hardware security module (HSM)** and 150 Always Free vault secrets)

Keep in mind that the resources listed here are available to your tenancy without incurring any costs, which is why it is categorized as always free. In addition, you can create IAM resources such as policies and networking (VCN, subnets, NSGs, and security lists) at no cost in order to support your application. For complete details of the **Always Free** resources, you can refer to the Oracle documentation.

I recommend checking for your **Always Free** resources in the Console under **Tenancy Service Limits**. You can check these by logging in to your cloud tenancy and opening the navigation menu. Under **Governance and Administration**, click on **Limits, Quotas, and Usage**. On this page, you can filter by services and region to see a list of resources available to you. *Figure 2.1* shows this navigation:

Figure 2.1 – Checking Always Free resources in the console

By now, you should have a high level of understanding of what Free tier and Always Free resources are available to you for the purposes of exploring OCI. Now we will look at how to go about provisioning them for learning purposes.

Always Free ADBs

OCI provides an Always Free ADB for any user associated with either the Free tier tenancy or paid tenancies. You can have two free ADBs in your tenancy. This could be your learning platform to understand features or even stand up a small production workload for your application development. Keep in mind that you can only provision ADBs in your home region. Also, the database version could be either 19c or 21c depending on your home region selection. Since not all regions support Always Free ADB, proper checking needs to be done while choosing the home region. Oracle takes care of migrating your free ADB to the latest version based on its schedule plan and notifies you in advance.

> **Quick note**
> Keep in mind that if your ADB is not in use for 7 consecutive days, Oracle will stop the ADB instance. You will not lose any data; you will have to restart the database if you want to continue using it. Another thing to keep in mind is that if there is inactivity for more than 7 days, Oracle puts the database on a deletion schedule after 90 days of continuous inactivity.

Oracle provides you with 1 **Oracle CPU processor** (**OCPU**) with 20 GB of storage for the database. In terms of workload types, you have the flexibility to select all four workload types: ATP, ADW, JSON, or APEX. Keep in mind that you have APEX support built in with ATP, ADW, and JSON as well.

Creating an Always Free ADB

To create this database, you need to have your user login and password. Once you authenticate to your cloud tenancy, you will be provided with the default dashboard, allowing you to carry out various tasks.

Open the the **OCI console** menu. Select **Oracle Database**, which will put **All Oracle Database** services on the right side of the dashboard. Select **Autonomous Database** from there, which will take you to the **Create Autonomous Database** page, which allows the creation of all ADB types. Click **Create Autonomous Database** and select the details for the ADB to be provisioned.

On the **Create Autonomous Database** screen, as shown in *Figure 2.2*, you need to understand certain details, such as compartment, database name, and details. Let's take a look at these details:

Figure 2.2 – ADB creation page in the console

Basic information for ADBs

In the **Provide basic information for the Autonomous Database** section shown in *Figure 2.3*, you need to provide the following details:

- **Compartment**: Create an ADB page that will have pre-populated names for your compartment, which should be changed to the name where you want to get this service provisioned. By default, it might be pointing to the root compartment of the tenancy.

- **Display name**: You can provide a name in this field as per your requirements. Certain organizations will have a naming convention already defined for any cloud resources they create in their tenancy. **Display name** helps you identify database resources. It can be any name you choose and does not need to be unique in your tenancy.

- **Database name**: You need to follow the rule enforced by OCI for database names. It must start with letters and can contain numbers, and it should not exceed 14 characters.

The following screenshot shows how to provide these details in the console:

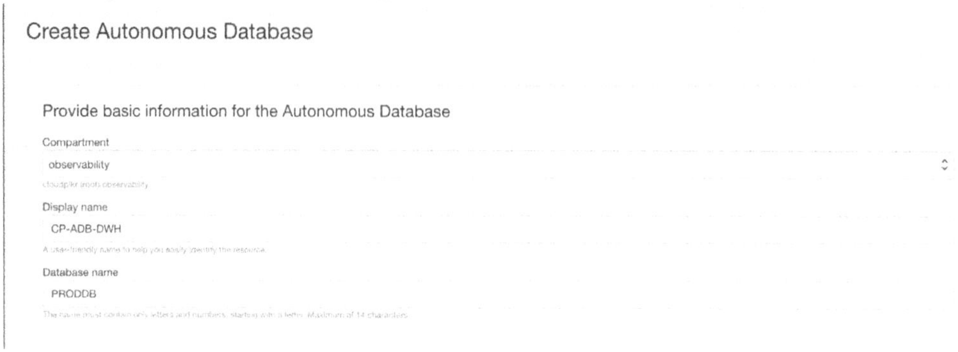

Create Autonomous Database

Provide basic information for the Autonomous Database

Compartment

observability

cloudpkr.iroot.observability

Display name

CP-ADB-DWH

A user-friendly name to help you easily identify this resource.

Database name

PRODDB

The name must contain only letters and numbers, starting with a letter. Maximum of 14 characters.

Figure 2.3 – ADB creation page – Basic information

Once you have filled in the basic information for your ADB, you can navigate to the next section, where you will be selecting your workload type.

Workload type for your ADB

This section allows you to choose your intended workload type. The options are **Data Warehouse**, **Transaction Processing**, **JSON**, and **APEX**. We discussed these types in the previous chapter. We will leave our selection at the default, **Data Warehouse**, as shown in *Figure 2.4*, but you should choose the most suitable one for your situation.

Figure 2.4 – ADB creation page – Selecting the workload type

Deployment type for your ADB

This section allows you to make your infrastructure selection for your intended workload type. The applicable types are **Shared Infrastructure** and **Dedicated Infrastructure**. In an Always Free tenancy, you cannot choose **Dedicated Infrastructure** because it is only available in the paid subscription. *Figure 2.5* shows this option during ADB creation:

Figure 2.5 – ADB creation page – Selecting the deployment type

The default option is **Shared Infrastructure**; we will leave it as is and move to the next section.

Configuration options for your ADB

This section allows you to move the selector to show the **Always Free** database configuration option. Since we are interested in deploying the **Always Free** option, move the selector to the right if not already. You need to make a selection for your database version between 19c and 21c (as of the time of writing this chapter). You can also see that you cannot change **OCPU count** and **Storage**, which is limited to 1 OCPU and 0.02 TB for Always Free ADB.

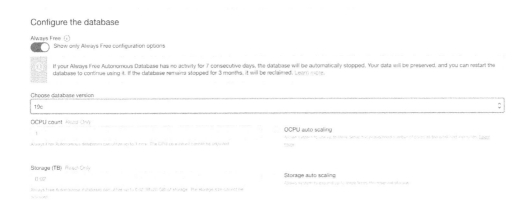

Figure 2.6 – ADB creation page – Database configuration options

In the preceding screenshot, you can see that **OCPU auto scaling** and **Storage auto scaling** are disabled. This is because of the limitations of Always Free ADBs.

Create administrator credentials for your ADB

The section allows you to set a password for the **ADMIN** user of your ADB. Once the database is provisioned, you will need this password to access database client tools such as SQL client, SQL Developer, and APEX tools. The following screenshot shows this option. Notice the username is grayed out, which means you cannot change it.

Figure 2.7 – ADB creation page – Setting up an ADMIN password

You need to make sure you comply with the password complexity requirements: 12 to 30 characters in length, including lowercase and uppercase letters, 1 number, and special characters, excluding quotation marks and strings containing admin.

Defining network access for your ADB

This section allows you to choose between several options for network access, such as **Secure access from everywhere**, **Secure access from allowed IPs and VCNs only**, and **Private endpoint access only**, which is more restrictive in terms of allowing only from within an OCI cloud tenancy and VCN. I have selected the second option and whitelisted my IP address for access, as shown in *Figure 2.8*:

Figure 2.8 – ADB creation page – Choosing network access type

Selecting the license type for your ADB

With an Always Free ADB, by default, this field is set to **License Included** and it cannot be changed. Additionally, you can provide your email address, which will allow notifications such as related to provisioning or maintenance to your instance.

In addition to the notification option provided here, you can use the OCI Events Service, which sends OCI resource-related events/alarms to your choice of tool. By using the OCI Events Service and the OCI Notifications Service, you can send a JSON payload related to the event to subscribed criteria in the notification section. A few examples include integration with email, Slack notifications, PagerDuty, and HTTPS notifications through webhooks for alerts if a database is going through a critical issue, has stopped, has been terminated, or has been patched.

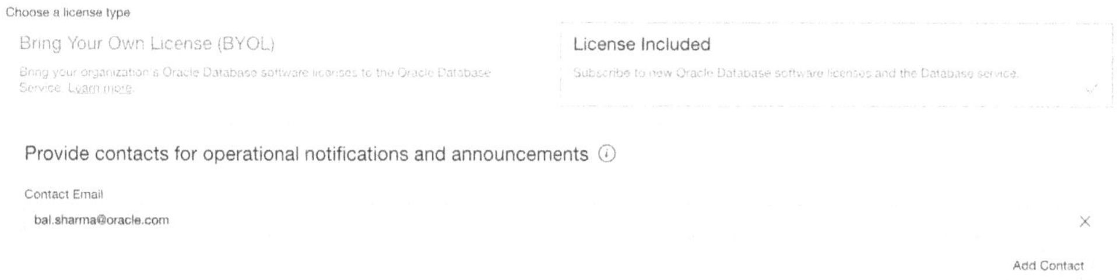

Choose a license type

Bring Your Own License (BYOL)

Bring your organization's Oracle Database software licenses to the Oracle Database Service. Learn more.

License Included

Subscribe to new Oracle Database software licenses and the Database service.

Provide contacts for operational notifications and announcements ⓘ

Contact Email

bal.sharma@oracle.com   ✕

Add Contact

Figure 2.9 – ADB creation page – License type for your ADB

As shown in the preceding screenshot, you can go ahead and select the appropriate licensing type, such as **Bring Your Own License (BYOL)** or **License Included**. For the Free tier, you have to go with the latter option. Selection for notification will allow you to get notified of important events.

Advance options for your ADB

This section allows you to define encryption keys. Oracle provides the flexibility to bring your own keys or use Oracle-managed keys. Additionally, you can provide any tags for your deployment and look at the maintenance schedule. You can provide free-form tags or user-defined tags. For user-defined tags, you need to have the necessary permission in the cloud tenancy. *Figure 2.10* shows the encryption options for you to select. You need to pay attention to them as they are related to security and operational aspects such as manageability.

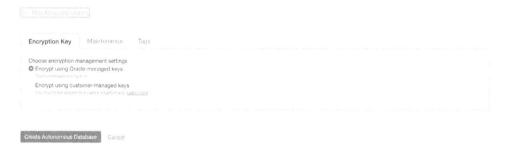

Figure 2.10 – ADB creation page – Advanced options for your ADB

This completes all the sections on the database creation page for your ADB. Once you hit the **Create Autonomous Database** button at the bottom of the page, Oracle will start provisioning the database instance for you, and it will be available in a few seconds to a minute. *Figure 2.11* shows the provisioning status in yellow while your instance is provisioning:

Figure 2.11 – ADB creation page – Database provisioning progress

Figure 2.12 shows the provisioning completion status in green. Now you are ready to explore your database instance and all the available options from within the console.

Figure 2.12 – ADB creation page – Database provisioning is complete

The preceding screenshot provides general information related to your selections earlier and several additional options for your exploration, such as backups, monitoring metrics, and service consoles for bringing more tooling into your ADB.

In order to use all the features, you will need to upgrade to a paid offering.

Networking, IAM, and security for shared and dedicated deployments

This section discusses important steps you need to perform, such as networking, IAM, and security through policies and permission of resources. They are applicable to any ADBs, such as shared or dedicated deployments you plan to do in OCI. Understanding these topics is critical because IAM and networking could be complex. Consider a scenario of identity federation where you need to integrate with another service provider (Microsoft Azure AD, Okta, etc.) for identity-related tasks. The same is true for networking. Imagine your databases need to be accessed only from within your cloud tenancy or from within a private subnet.

Prerequisites – IAM considerations

OCI provides IAM integration as a shared security service across all resources deployed in a tenancy. You can consider it as a common framework that helps integrate OCI IAM for resource authentication and authorization needs applicable to all the integration methods using the Console, CLI, SDK, or REST interfaces.

You might need to connect with your administrator or IAM personnel who will be setting up IAM resources such as groups, compartments, policies, dynamic groups, and federations in order to control user behavior when accessing cloud resources based on the user profile. Consider various tasks that a DBA might need to perform, such as creating or granting access to a new database user (such as a developer), creating new databases, using network resources such as **Network Security Groups** (**NSGs**), opening a port through an NSG, whitelisting IP addresses for access, or network resource access such as a VCN. As a DBA, you need access to a backup location, permission to list or manage backups, download files, and more. All these tasks are controlled through policies defined in your cloud tenancy. Normally, an infrastructure administrator sets up a profile for you and provides details such as compartment and network resources to be used for your tasks.

Creating an ADB requires certain privileges using a policy statement defined by your administrator irrespective of the access method used, such as the console, REST, OCI CLI, and SDKs. If you encounter a message stating that you don't have permission to create resources or an unauthorized warning message, you should get in touch with your infrastructure administrator.

Policy requirements for your ADB

You need to understand a few concepts relating to policy requirements. To start with, we will discuss aggregate resource types and individual resource types in the context of ADBs. Consider an aggregate resource type as a collection of individual resource types that directly follow policies defined at the aggregate level. This aggregate resource type is *autonomous-database-family*. If we define a policy at this level, it translates to four separate policies for the group: *autonomous-databases*, *autonomous-backups*, *autonomous-container-databases* (applicable to only the dedicated Exadata Infrastructure option for ADBs), and *cloud-autonomous-vmclusters* (applicable to only the dedicated Exadata Infrastructure option for ADB).

Quick note

There are two groups of resource types in ADB. The aggregate resource type is autonomous-database-family. Individual resource types are autonomous-databases, autonomous-backups, autonomous-container-databases (dedicated Exadata infrastructure only), cloud-autonomous-vmclusters (dedicated Exadata infrastructure only), and database-connections.

Resource types for a dedicated ADB

For dedicated ADB provisioning, there are two individual resource types, which we just covered. The aggregate resource type for our deployment is served by `database-family`. This aggregate class covers several individual resource types: `cloud-exadata-infrastructures`, `cloud-vmclusters`, `db-nodes`, `db-homes`, `databases pluggable-databases`, and `db-backups`.

Supported variables

All general variables are supported; as you go from **inspect|read|use|manage**, the level of access gets cumulative. For provisioning, you can use the `target.workload` type variable, with which you can define workload types such as OLTP, DW, AJD, and APEX for all four workload requirements.

Let's take a look at an example that shows how a policy can be written to grant manage permission to a group called `ADB-Admins` based on target workload type:

```
Allow group ADB-Admins to manage autonomous-database in tenancy where
target.workloadType = 'workload_type'.
```

The resource family covered by `autonomous-database-family` can be used to grant access to database resources associated with all the ADB workload types.

ADB permission and API coverage

The resource family covered by the `autonomous-database-family` resource type can be used to grant access to database resources associated with all the ADB workload types. The following table provides details of the permissions and API coverage for the `autonomous-databases` resource type. The same resource type is used to grant access to the ADB resources used by either ADW or ATP.

For the `autonomous-database` family, the following table provides details about APIs and their functionality in terms of coverage. As you can see, permission can vary based on the keywords **inspect**, **read**, **use**, and **manage**.

Verbs	Permissions	Fully Covered API	Partially Covered API

inspect	`autonomous-database-in-spect`	`ListAutonomous Databases`
		
		`GetAutonomous Database`
			NA

read	`autonomous-database-content_read`	NA	`CreateAutonomous DatabaseBackup` (needs manage autonomous-backups in addition)
use	`autonomous-database-content_write,` `autonomous-database-update`	`UpdateAutonomous Database`	`RestoreAutonomous-Database` (needs read autonomous-backups in addition) `ChangeAutonomous DatabaseCompartment` (needs read autonomous-backups in addition)
manage	`autonomous-database-create,` `autonomous-database-delete`	`CreateAutonomousDa-tabase`	NA

For the `autonomous-backups` family, the following table provides API details and their functionality in terms of coverage. These operations provide details on backup topics such as the ability to list, use, read, and manage backups for your autonomous databases. Permission levels can vary based on the keywords **inspect**, **read**, **use**, and **manage**, from lowest privilege to highest privilege.

Verbs	Permissions	Fully Covered API	Partially Covered API
inspect	`autonomous-db-backup-inspect`	`ListAutonomous DatabaseBackups,` `GetAutonomous DatabaseBackup`	NA
read	`autonomous-db-backup-content-read`	NA	`RestoreAutonomousDatabase` (needs use `autonomous-databases` in addition) `ChangeAutonomousDatabase Compartment` (needs use `autonomous-databases` in addition)
use	NA	NA	NA
manage	`autonomous-db-backup-create,` `autonomous-db-backup-delete`	`DeleteAutonomous DatabaseBackup`	`CreateAutonomousDatabase Backup` (needs read `autonomous-databases` in addition)

The `autonomous-container-databases` (used in dedicated ADBs) family provides APIs to manage container databases on ADBs. These databases are dedicated instances and can be managed through various API calls such as **list**, **read**, **use**, and **manage**.

Verbs	Permissions	Fully Covered API	Partially Covered API
inspect	`autonomous-container-database-inspect`	`ListAutonomous ContainerDatabases, GetAutonomous ContainerDatabase`	NA
read	NA	NA	NA
use	`autonomous-container-database-update`	`UpdateAutonomous ContainerDatabase` `ChangeAutonomous ContainerDatabase Compartment`	`CreateAutonomousDatabase` (also needs `manage autonomous-databases`)
manage	`autonomous-container-database-create` `autonomous-container-database-delete`	NA	`CreateAutonomousContainer Database, Terminate AutonomousContainerDatabase` (both also need `use cloud-autonomous-vmclusters,` `use cloud-exadata-infrastructures`)

`cloud-autonomous-vmclusters` (used for dedicated ADBs) is the class used to provide access to VM clusters defined on dedicated Exadata nodes. It covers permission ranging inspect to manage in order to control access.

Verbs	Permissions	Fully Covered API	Partially Covered API

inspect	cloud-autonomous-vm-cluster-inspect	`ListCloud AutonomousVmClusters`
		
		`GetCloud AutonomousVmCluster`
			NA

read	NA	NA	NA

use	cloud-autonomous-vm-cluster-update	`UpdateCloudAutonomous VmCluster`
		
		`ChangeCloudAutonomous VmClusterCompartment`
			`CreateAutonomous Database`
			
			needs manage `autonomous-databases`
			
			`CreateAutonomous ContainerDatabase` -
			needs manage `autonomous-container-databases`

manage	`cluster-autonomous-vm-cluster-create`
	
	`cluster-autonomous-vm-cluster-delete`
		NA	`CreateCloud AutonomousVmCluster, DeleteCloud Autonomous VmCluster`
			
			-Above API calls also need use `vnics`, use `subnets`, use `cloud-exadata-infrastructures` permissions.

ADB – List of APIs

The `Autonomous Exadata Infrastructure` resource class facilitates infrastructure admin to manage and control resources required for provisioning the ADBs. Permissions control the level of activities administrators can do inside the cloud tenancy. With the manage permission, authorized users can create ADB resources, modify them, or even terminate them.

API	Operation
LaunchAutonomous ExadataInfrastructure	Creates an Autonomous Exadata Infrastructure resource
UpdateAutonomous ExadataInfrastructure	Configures automatic maintenance schedule for the Autonomous Exadata Infrastructure resource
GetMaintenanceRun	Displays the details of a maintenance run that is scheduled, in progress, or that has ended
ListMaintenanceRun	Gets a list of maintenance runs in a specified compartment
ChangeAutonomous ExadataInfrastructure Compartment	Moves an Autonomous Exadata Infrastructure resource to another compartment
TerminateAutonomous ExadataInfrastructure	Deletes an Autonomous Exadata Infrastructure resource

The `Autonomous Exadata VM Cluster` resource class provides the required API for creating, deleting, and updating a VM cluster, as shown here:

API	Operation
CreateAutonomousVmCluster	Creates an Autonomous VM cluster
ListAutonomousVmClusters	Gets a list of Autonomous VM clusters in the specified compartment
GetAutonomousVmCluster	Gets information about the specified Autonomous VM cluster
UpdateAutonomousVmCluster	Updates the specified Autonomous VM cluster

API	Operation
DeleteAutonomousVmCluster	Deletes the specified Autonomous VM cluster
ChangeAutonomousVmClusterCompartment	Moves an Autonomous VM cluster and its dependent resources to another compartment

ADB admins use the Autonomous Container Database resource class APIs to perform operations such as creating container databases, controlling backup and patching operations, changing container database compartments, and even terminating ADBs. The following table lists the APIs for administrators to use:

API	Operation
CreateAutonomousContainerDatabase	Creates an autonomous container database.
UpdateAutonomousContainerDatabase	Sets the backup retention period for an autonomous container database. Sets the maintenance patching type of an autonomous container database.
UpdateMaintenanceRun	Skips a container database maintenance run.
ListMaintenanceRun	Gets a list of maintenance runs in a specified compartment. Can be used to see maintenance history and scheduled maintenance runs.
RestartAutonomousContainerDatabase	Performs a rolling restart on a container database.
ChangeAutonomousContainer DatabaseCompartment	Moves a container database to another compartment.
TerminateAutonomousContainerDatabase	Terminates a container database.

The `Autonomous Database` resource class is used mainly by database administrators to control activities such as creating PDBs, updating any attributes related to PDBs, and controlling access. The following table lists the APIs for managing ADB operations in OCI:

API	Operation
CreateAutonomousDatabase	Creates ADBs of either the transaction processing or warehouse workload types
ListAutonomousDatabases	Gets a list of ADBs
GetAutonomousDatabase	Gets the details of the specified ADB
UpdateAutonomousDatabase	Updates one or more attributes of the specified ADB
ChangeAutonomousDatabaseCompartment	Moves the ADB and its dependent resources to the specified compartment
StartAutonomousDatabase	Starts the specified ADB
RestartAutonomousDatabase	Restarts the specified ADB
StopAutonomousDatabase	Stops the specified ADB
DeleteAutonomousDatabase	Deletes the specified ADB
GenerateAutonomousDatabaseWallet	Downloads the client credentials for an ADB
UpdateAutonomousDatabaseWalletDetails	Rotates the wallet for an ADB
AutonomousDatabase	Gets the access URLs for APEX and SQL Developer Web

The `Managing Backup Destinations` resource class facilitates backup operations such as creation and listing information. The following table lists all the available APIs:

API	Operation
CreateBackupDestination	Creates a backup destination
DeleteBackupDestination	Deletes a backup destination
GetBackupDestination	Gets information about the specified backup destination
ListBackupDestination	Gets a list of backup destinations in the specified compartment

API	Operation
UpdateBackupDestination	If no database is associated with the backup destination: For a RECOVERY_APPLIANCE backup destination, updates the connection string and/or the list of VPC users For an NFS backup destination, updates the NFS location

The Backing Up an Autonomous Database Manually resource class can be used to perform manual backup activities on ADBs:

API	Operation
ListAutonomousDatabaseBackups	Gets a list of ADB backups based on either the autonomousDatabaseId or compartmentId specified as a query parameter
GetAutonomousDatabaseBackup	Gets information about the specified ADB backup
CreateAutonomousDatabaseBackup	Creates a new ADB backup for the specified database based on the provided request parameters

The Restore Autonomous Database resource class is used for database restoration tasks and for facility administrators to use a backup previously taken for restoration purposes:

API	Operation
RestoreAutonomousDatabase	Restores an ADB from a backup

The cloning an API is used to create another ADB instance for purposes such as testing or development:

API	Operation
CloneAutonomousDatabase	Clones an ADB

Networking considerations

Networking plays a great role in accessing databases within a customer tenancy. Applications need to access databases securely and don't want to send network traffic through the public internet. When ADBs were launched as a shared offering, the only way to access them was through a public endpoint. A private endpoint for a shared ADB allows you to take advantage of your **Virtual Cloud Network** (**VCN**) and create a private endpoint for database access. By doing this, it blocks database access from public endpoints. Another consideration for a private endpoint is that to simplify connections across a different network, such as accessing on-premises to the cloud, you don't need to set up transit routing through your VCN.

This provides flexibility and control as you can restrict traffic flow using a security list and NSGs within the VCN for your ingress and egress needs.

Steps to configure a private endpoint for a shared ADB

There are certain requirements when you are configuring a private endpoint, such as IAM policies and subnet-specific configurations.

IAM policies control who can do what on the ADB. Any access, whether through the console or CLI, REST, or an SDK, is controlled through policies. There are specific policies that control private endpoints.

Private endpoint configuration with ADBs requires an IAM policy statement to be configured by admins. When deploying an ADB, you need to have to manage permission with a VCN in order to use a PE.

Cloud Operation	IAM Policy required
private endpoint configuration	*use* `vcns/subnets/network-security-groups` can be used at the compartment level where VCN resides *manage* `private-ips/vnics` can be used at compartment level where VCN resides, or database needs to be provisioned

Once you have IAM policies in place, you can decide on the VCN and subnets where you want to create private endpoints. They could be a new VCN and subnet or existing ones. While creating a subnet, you need to keep in mind that the VCN should be configured with default DHCP options. Optionally, you can create an NSG and specify rules for connecting to the shared ADB.

Most enterprises have strict security requirements, and they need a private endpoint that provides a private IP address, and any internet access to the data, either at the bucket or namespace level, is blocked. Restricting access to the selected network allows all communication to happen from either one or a set of VCNs, and services can only accept communication from allowed VCNs and subnets in the tenancy. This helps to prevent any kind of data exfiltration for the customer.

For detailed information on private endpoints, please refer to *Chapter 7*.

Deploying a serverless shared ADB

Deploying a serverless shared ADB allows you to create all four kinds of database (ADW, ATP, AJD, and APEX). OCI provides per-second billing, which means provisioned OCPU and storage consumption are billed by the second with a minimum of 1 min usage period.

In the previous section, when we discussed deploying a Free tier ADB, you saw all the prerequisites and steps. Deploying a shared ADB will follow the same steps.

Deploying a dedicated Exadata infrastructure

So far, you have learned how to deploy a shared ADB. Now, we will see how a dedicated ADB is provisioned. A dedicated ADB is provisioned on a dedicated Exadata infrastructure through the OCI Console, via an API, or through Terraform. The ADB can be of either type as supported by the shared infrastructure or by a standby database type for disaster recovery. To create any type of database, you need to first provision the infrastructure and an Autonomous Container Database.

Steps to provision a dedicated ADB

In order to provision a dedicated ADB, you need to have fleet administrator privilege. After logging into the OCI Console, open the **OCI console** menu. Select **Oracle Database**, which will open **All Oracle Database services** on your home screen. Select **Autonomous Database**, which will open the **Create Autonomous Database** page for you, which allows the creation of all ADB types. Click **Create Autonomous Database** and select the details of the ADB to be provisioned.

Create Exadata infrastructure for a dedicated ADB

You need to provide basic information for the Exadata infrastructure, such as compartment, display name, availability domain, the shape of the machine, and compute and storage configuration. Please see the following screenshot for the details:

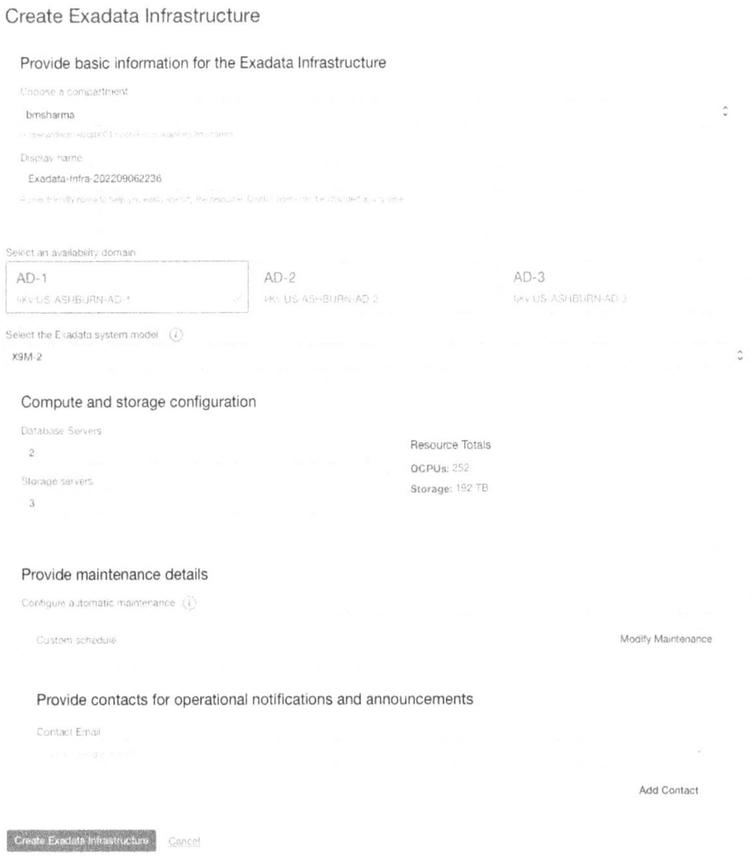

Create Exadata Infrastructure

Provide basic information for the Exadata Infrastructure

Choose a compartment

bmsharma

Display name

Exadata-Infra-202209062236

Select an availability domain

AD-1	AD-2	AD-3
VKv:US-ASHBURN-AD-1	PKt:US-ASHBURN-AD-2	tXv:US-ASHBURN-AD-3

Select the Exadata system model

X9M-2

Compute and storage configuration

Database Servers

2

Storage servers

3

Resource Totals

OCPUs: 252

Storage: 192 TB

Provide maintenance details

Configure automatic maintenance

Custom schedule   Modify Maintenance

Provide contacts for operational notifications and announcements

Contact Email

Add Contact

Create Exadata Infrastructure   Cancel

Figure 2.13 – Creation of an Exadata infrastructure for a dedicated ADB

Creating an Exadata VM cluster to be used for a dedicated ADB

In this section, you need to provide the following details:

- **Compartment**: Create an ADB page that will have pre-populated names for your compartment, which should be changed to the name where you want to get this service provisioned. By default, it might be pointing to the root compartment of the tenancy.

- **Display name**: You can provide any name in this field that you want. Certain organizations will have a naming convention for any cloud resources they create in their tenancy. Display names help you identify database resources. This can be any name you choose and does not need to be unique in the tenancy.

- **Database name**: You need to follow the rule enforced by OCI for naming a database. It must start with letters and can contain numbers, and should not exceed 14 characters.

A VM cluster for Exadata resides on all compute nodes, which makes a cluster and depends on the shape of the dedicated Exadata hardware selected (such as a quarter rack or half rack). The purpose of multiple VMs in a cluster is to provide high availability for your application databases. All the resources from the VM cluster are available for your database workload consumption. Let's navigate to the ADB option in the console, as shown in *Figure 2.14*, to create the VM cluster we talked about.

Figure 2.14 – Diagram showing the Console screen for Exadata VM cluster creation

Click on **Create Autonomous Exadata VM Cluster** and provide details such as the compartment and the display name, and select the compartments where you have provisioned Exadata Infrastructure:

Figure 2.15 – Diagram showing details for Exadata VM cluster creation

You need to select the Exadata infrastructure you created for dedicated deployment in your tenancy in the previous step.

Creating an autonomous container database

In this section, you need to provide the same details as with the deployment of a shared ADB. Please refer to that section for a general understanding of the compartment, display name, and database name.

The following screenshot shows how to navigate and create the container database:

Figure 2.16 – The navigation page for creating an autonomous container database

On this screen, click on **Create Autonomous Container Database** and fill out the details as shown in the following screenshot. For basic information such as compartment and display name, you need to pick the Exadata infrastructure and VM cluster you created earlier. It also provides the ability to create the data guard configuration at the time of provisioning.

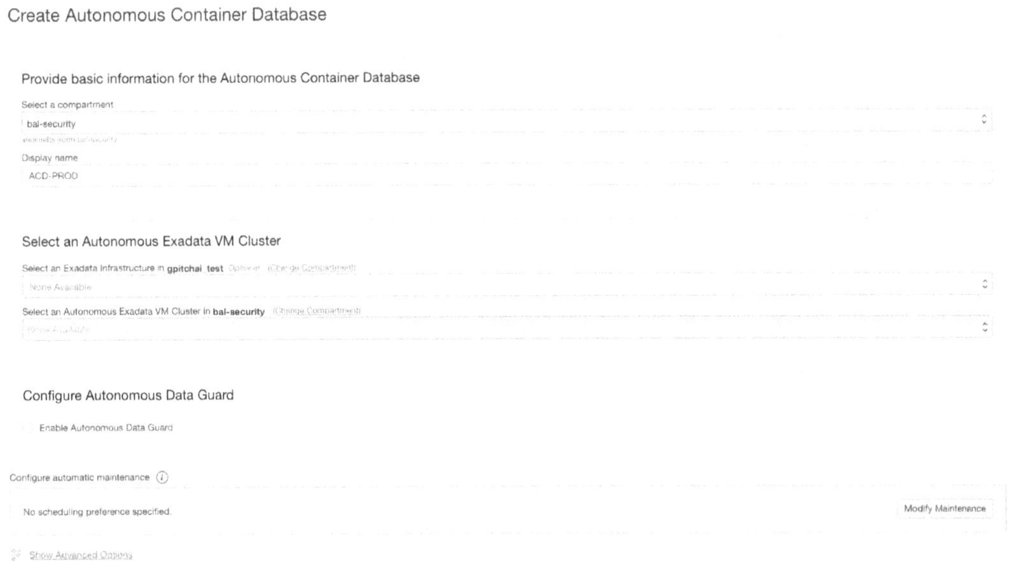

Figure 2.17 – Creating an autonomous container database

If you plan to create a data guard configuration, you need to have created a peer Exadata Infrastructure along with VM clusters. You can decide on a protection mode for the configuration. *Figure 2.18* shows detailed inputs for this configuration.

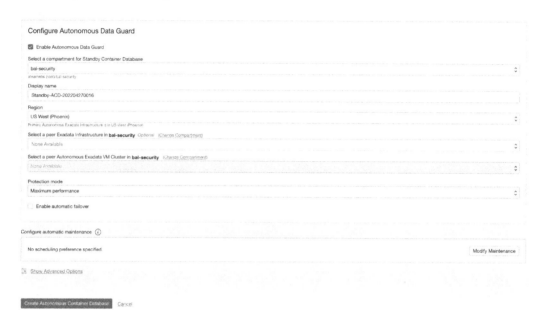

Figure 2.18 – Data Guard configuration for the autonomous container database

After providing details, click on **Create Autonomous Container Database** and the display will show the progress and the database's state once provisioning is complete.

If you've followed these steps, you have provisioned an autonomous container database with Autonomous Data Guard enabled. As a DBA, you can create a standby database for your autonomous container database with all ADBs created automatically.

Summary

This chapter explains various provisioning aspects, IAM, and networking considerations for ADBs. It also explains how you can take advantage of a Free tier cloud subscription to get familiar with OCI and ADBs. You should be able to design different roles needed to deploy and maintain your ADB and apply the principle of least privilege. A secure networking consideration such as private endpoints enhances the overall security posture of your database deployment. We have discussed ADBs and their deployment options. Let's talk about how to migrate to an ADB in the next chapter.

Part 2 – Migration and High Availability with Autonomous Database

Let's learn how you can migrate existing databases – be it an on-premises transactional database or a large data warehouse on Autonomous Database in OCI, you should be able to clearly understand the best method of migrating databases depending on the service level of agreement for existing applications and their downtime requirements. Once databases are migrated to Autonomous Database, cloud administrators can design high availability options, such as standby databases and replicas using the cloning functionality, as well as switching over and failing over databases with the Autonomous Data Guard capabilities.

This part comprises the following main chapters:

- *Chapter 3, Migration to Autonomous Database*
- *Chapter 4, ADB Disaster Protection with Autonomous Data Guard*
- *Chapter 5, Backup and Restore with Autonomous Database in OCI*
- *Chapter 6, Managing Autonomous Databases*

3
Migration to Autonomous Database

With some understanding gained about what exactly an autonomous database service offering is in Oracle Cloud Infrastructure, as discussed in *Chapter 1, Introduction to Oracle's Autonomous Databases,* , we will now set the stage in this chapter to enable you to simplify your migration approach to an autonomous database service. Throughout this chapter, we will discuss at a high level what options are available for you to migrate to an autonomous database, understand the restrictions currently existing in the service, and then deep dive into one of the simplified tools available that can automate your migration to an autonomous database service. By the end of this chapter, you will have obtained a detailed understanding of how this automation works, enabling you to put this into practice when migrating one or more databases to an autonomous database, irrespective of the type **Autonomous Data Warehouse (ADW)/Autonomous Transaction Processing (ATP)** database, with less manual effort. So, let's start by taking into account some key considerations.

We will cover the following main topics in this chapter:

- Migrating to autonomous methods
- Online and continuous data migration
- Test master creation

Let's begin!

Migration considerations

Oracle Autonomous Database has been designed to be fast, easy, and secure, with the objective of overcoming most of the management overheads usually associated with any database. We learned about these capabilities in *Chapter 1, Introduction to Oracle's Autonomous Databases*. Now, the question is, can your application database be hosted on an autonomous database so that you can take advantage of autonomous capabilities? Well, the answer is yes in most cases. However, there is also a word of caution.

Autonomous Database, although it is an Oracle database, does have certain restrictions compared to a usual enterprise Oracle database, and it is worth reviewing the limitations once by clicking on the following link and reviewing the details to see what features are not supported:

```
https://docs.oracle.com/en/cloud/paas/autonomous-database/adbsa/
experienced-database-users.html#GUID-58EE6599-6DB4-4F8E-816D-
0422377857E5
```

These limitations exist for one or more reasons to make the database more secure, fast, and efficient in performance, taking away a large share of the usual database manageability or administration overhead that you would otherwise be involved in, allowing you to focus more on your business operations.

For example, you cannot create tablespaces in Autonomous Database, you cannot alter certain initialization parameters, you don't need to create indexes on tables, and you can leave query optimization to Autonomous Database, as it can apply the best query optimization technique to give you the best query response time of any database on the market today, with exactly the same workload type. However, noting the restrictions is equally important in order to make the right decision when planning your migration to the Autonomous Database platform. Additionally, you must perform validation of applications in a non-production environment to rule out any incompatibility and resolve them with minimal changes. Nevertheless, Autonomous Database has been certified with many Oracle and third-party data integration and analytical tools, and it is being widely adopted by several customers to meet their analytical or transactional database requirements in a public cloud or Cloud@Customer.

Once a decision has been made to migrate to Autonomous Database, choosing the right migration approach is very crucial. Oracle provides a **Cloud Migration Advisor** tool that suggests the optimal method for migrating to Autonomous Database based on the Oracle source database, with releases starting from 11g. It also suggests a method that can be used when migrating to an autonomous data warehouse from **AWS** Oracle database deployments on EC2, RDS, or Redshift. Take a look at the Cloud Migration Advisor by clicking on the following link, and play around by selecting the source version and target you want to migrate to:

```
https://www.oracle.com/webfolder/s/assets/webtool/cloud-migration-
advisor/index.html
```

As a recommendation, I would suggest that you first check the advisor while planning your migration to Autonomous Database. As an example, suppose I wish to migrate an 11g database to an autonomous transaction processing database; the Cloud Migration Advisor would suggest different approaches, such as **Zero Downtime Migration (ZDM)**, **OCI Data Migration Service (DMS)**, and the Database Migration Workbench in Oracle Enterprise Manager 13.4 Release Update 7 and above, which can automate your migration, in addition to convention tools/utilities such as Oracle GoldenGate, Data Pump, SQL Developer, and SQL *Loader that you can adopt, as depicted in *Figure 3.1*.

11g,12c,18c,19c,21c

ZDM(Oracle Golden Gate, datapump)
OCI Data Migration Service (DMS)
OEM Migration workbench
Others (Datapump, SqlDeveloper etc)

Oracle Autonomous Database

Figure 3.1: Migration approaches

Database migration is a critical operation, and several considerations need to be made before eventually migrating the database. These could be resizing the target database based on the source database, the compatibility of your in-house application with the target database, the needful configuration of compute resources that must deliver better performance in comparison to your source system, target platform support, and so on. The advantage of Autonomous Database is that it offloads the majority of these concerns when you are planning to migrate to the autonomous database like what should set the initial target storage size so that migration does not fail, how should I choose the compute and appropriate memory size, the database-specific parameter configuration, performance characteristics, and so on. It is a self-tuning database optimized for the kind of workload while leveraging the high-performing Exadata infrastructure on which it is hosted. And with the auto scaling features for compute and storage, you need not worry about setting the initial sizing while migrating the database. Depending on the type of workload, you can choose to migrate to either an autonomous data warehouse or an autonomous transaction processing database. Each of them provides predefined service consumer groups that your applications can be configured to connect with, in order to meet the different levels of parallelism or concurrency requirements to run the different application queries.

When migrating Oracle databases to Autonomous Database, it is essential that all the incompatibilities are identified and resolved before the migration task is executed. Oracle provides a Java-based tool named the **Cloud Premigration Advisor Tool** (**CPAT**) that can automate this action. It can assess the source Oracle database and target Autonomous Database to perform a series of checks and determine the possibility of a successful migration to Autonomous Database. For example, it can perform checks at the schema level, it can check for tables that are using deprecated/unsupported features, or it can check for features that are not supported in Autonomous Database. It can even perform checks at the data level within the schema or just at the database instance level, such as identifying instance parameters not editable in Autonomous Database. You can also control the scope of whether the checks should run on all user database schemas or a selected list of schemas. The output of the tool is a report containing the results of the different checks – that is, whether the check passed, failed, a review is suggested, a review is required, or action is required. For each of the results, you get more details about the checks, and you can take appropriate action as necessary. The best part is that the tool is integrated into a ZDM solution, which is our main focus in this chapter. Database migration activity does incur some downtime, and depending on the method you adopt for migration, you can keep downtime to a minimum, but you must estimate that by performing a migration exercise on the non-production image of your database that closely resembles your production database size. Oracle does provide a near-zero downtime migration approach using Oracle GoldenGate, which reduces your application downtime to a bare minimum, with the downtime being proportional to the time needed to switch your application connection to the migrated autonomous database, syncing 100% with your source database. Discussing the migration approach with Oracle GoldenGate would be an elaborative subject, as you must have prior knowledge of it before we can discuss how it needs to be configured to achieve near-zero time migration; hence, it is beyond the scope of this book. This chapter is restricted to discussing one of the most recommended automated approaches for migration to Autonomous Database.

Migration to Autonomous Database can be carried out from an Oracle or non-Oracle database. Migration from Oracle databases is simplified, as there are already existing database utilities such as Data Pump and SQL *Loader that can be used. Migration from a non-Oracle database can also be done by using tools such as SQL Developer, although at the time of writing, only MSSQL or Sybase database migration support this tool. If your organization has certain third-party database migration tools that have a certified autonomous database, you may opt for those, but it is recommended to use SQL Developer. For databases that do not have direct support with any of the available migration tools, migration can be done with data integration tools such as Oracle Data Integrator or Oracle GoldenGate. For mission-critical applications with little to no downtime constraint, it is preferable to use Oracle GoldenGate, which is a near-zero downtime migration solution.

Some of the other approaches you can adopt are as follows:

- Use database links in the case of Oracle data sources (using the `datapump network_link` option), but this has great dependency on your network bandwidth; hence, you must be on a fast-connect network infrastructure to import faster.

- Take a Data Pump export of Oracle data, store it in OCI Object Storage, and import it using Data Pump Import, but this requires you to use Oracle client 18c (which is automated with the `mv2adb` package tool).

- Store your data in a delimited text file or even an Avro or Parquet format in Oracle Object Storage, and then use the `DBMS_CLOUD` APIs to load data directly into Autonomous Database.

- Store your data in a delimited format in OCI Object Storage and create an external table on the files to be read, and then load it into database tables.

So, that's some insight into how we can migrate our databases, which could be on-premises or on cloud database services such as Oracle DBCS, to Autonomous Database. Now, let's discuss one of the migration methods to migrate an on-premises database to Oracle Autonomous Database, which is called zero-downtime migration and is the recommended tool from Oracle to automate your entire database migration process while moving workloads to Oracle Cloud Infrastructure.

About the Zero-Downtime Migration tool

ZDM is Oracle's recommended solution to migrate Oracle databases on-premises, Oracle Cloud Infrastructure Classic, or a non-Oracle cloud into Oracle Cloud Infrastructure Database Services, which can be on virtual machines, bare-metal, an Exadata Cloud service, or the flagship Autonomous Database services. ZDM supports migration from 11.2.0.4, 12.1.0.2, 12.2.0.1, 18c, 19c, and 21c database releases to OCI database services. The current release, ZDM 21.3, supports cross-version migration, thus allowing you to perform an in-flight upgrade as part of the migration workflow. You can easily migrate from non-CDB databases to a pluggable database in one workflow instead of using a multi-level migration workflow. You can suspend, resume, abort, query status, modify, audit, add, or delete a running migration job, which can be performed either on a single database or a fleet of databases.

ZDM leverages Oracle Maximum Availability Architecture best practices that automate the entire migration process, reducing the chances of human error. It is robust, flexible, resumable and under the hood, using proven technologies that include Data Pump, Oracle Data Guard, Oracle GoldenGate, and Oracle RMAN, giving you the capabilities for both an online and offline logical and physical migration, with zero-data loss assurance. ZDM is available as downloadable software that is installed on a host and runs as a service. Post-installation, it provides all the CLIs needed to automate migration, which we will see in this chapter as we migrate an on-premises database to Autonomous Database. Currently, it is only supported on the Linux platform – that is, Oracle Linux 7.

This is just a brief introduction to zero-downtime migration, and I would recommend checking the public documentation on zero-downtime migration for more details.

Migrating on-premises database to Autonomous Database using ZDM

As discussed briefly in the previous section, ZDM offers different methods to automate migration to Autonomous Database, and which method you choose depends upon the source and target Oracle database, the version, and the amount of downtime that your organization can afford. At a high level, here are the four different approaches that can be used for migration:

- Physical online migration
- Physical offline migration
- Logical online migration
- Logical offline migration

In this section, we are going to learn about the ZDM logical online migration method for migrating from an Oracle database to Autonomous Database. We will discuss the different steps and configurations needed to completely automate the migration of an on-premises transaction database to an Oracle autonomous transaction-processing database in Oracle Cloud Infrastructure, with the ZDM tool.

The overall architecture typically would work in the manner shown in *Figure 3.2*. The ZDM service is the central entity that orchestrates the entire migration process from the source database to the target database, which in this case is an Oracle autonomous transaction-processing database. This migration process involves various components in Oracle Cloud Infrastructure, which includes Oracle GoldenGate and OCI Object Storage – instrumental in completing the zero-downtime online migration to Autonomous Database.

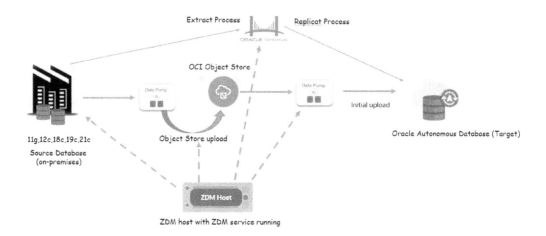

Figure 3.2: Logical online migration using ZDM – architecture

The migration workflow, as shown in *Figure 3.3*, is a series of steps that are automated by ZDM. Let's discuss the different steps involved while performing zero-downtime migration to Autonomous Database using ZDM.

Figure 3.3: Logical online migration using ZDM – workflow

Let's go through the workflow steps in the following list:

1. **Install and configure ZDM**: Designate a host to be used to configure the ZDM service. This host is called the ZDM host, configured with Oracle Linux 7. In this step, you download the ZDM software and configure the service. We will discuss this more in the forthcoming sections.

2. **Complete Prerequisites**: Perform some of the prerequisites needed, which could be SSH key generation, OCI CLI setup, Object Storage bucket creation, GGHUB setup, and so on.

3. **Prepare response file**: Use the response file template and modify the parameters required to perform migration into Autonomous Database from the source database.

4. **Dry-Run**: Invoke the ZDM CLI to first evaluate the response file, where it does a pre-check with the response file parameters to ensure that the configuration is correct to proceed with the migration. Once completed, invoke the ZDM CLI to initiate the migration job.

5. **Invoke zdmcli migration**: If the dry run of the `zdmcli` command returns COMPLETED for each step, then you invoke `zmdcli` to initiate the migration.

These are the high-level steps that are involved when performing migration using ZDM. This entirely automates your migration, with just one standard template file that has the inputs that ZDM needs to automate the migration.

Figure 3.4: ZDM migration steps

Figure 3.4 is a high-level representation of all the steps performed when `zdm` migration is invoked using `zdmcli`:

1. **Validation**: Here, ZDM will perform validation of the connectivity to the source database, the target database, the backup location, Data Pump settings, and the GoldenGate hub

2. **Configure GGHUB source extract**: Here, ZDM prepares the GGHUB for extracting active transactions from the source database before Data Pump is initiated.

3. **Data pump Export and transfer**: In this step, ZDM takes a Data Pump export of the source schemas and transfers them to the OCI Object Store bucket

4. **Data pump Import into target**: The dump that exists in the Object Store bucket is imported into Autonomous Database

5. **Synchronize Target Using GG replication**: Here, the replication process is prepared in GGHUB and the process is started to replicate all active transactions that were generated from the point when the Data Pump export was started, until they are imported into the target autonomous database. This ensures that data in Autonomous Database is in sync with the source database.

6. **Switch Application**: The implementer decides when to switch the application to point to the Autonomous Database instance, which would basically mean a change to the database connection entries provided the network connectivity between the application network and Autonomous Database virtual cloud network has been previously enabled and validated.

7. **Cleanup**: ZDM does an auto cleanup of the different resources it has created as part of the migration process.

At a high level, we have discussed the different steps involved in the entire end-to-end ZDM to Autonomous Database. In the following sections, we will learn about steps we must follow while performing actual database migration to Autonomous Database.

Implementing migration to Autonomous Database using ZDM

Here onward, we are going to follow the steps sequentially – that is, to be carried out actually while performing an actual migration of database schemas from an on-premises database to an Oracle autonomous transaction-processing database in OCI. The workflow would remain the same for an online logical ZDM into an Autonomous Database target, which may be on an OCI public cloud, Exadata Cloud@Customer, or Dedicated Region Cloud@Customer.

Installing and configuring ZDM software on a dedicated host

ZDM software is preferably set up on a host that is different from the database server; however, it can be shared for other purposes, but there should be no Oracle grid infrastructure running on the host. The supported OS platform is the Oracle Linux 7 platform, and it should have 100 GB of storage available. It should be possible to establish network communication between this host and the source and target databases. ZDM runs as a service, and this host is named the ZDM service host.

Before you can begin downloading and installing ZDM software, there are few prerequisites to be completed, as follows.

Installing the required packages:

You must install the `glibc-devel`, `libaio`, `expect`, and `Oraclelinux-developer--release-el7` packages:

```
[root@zdmhost ~]# yum list expect
[root@zdmhost ~]# yum list libaio*
[root@zdmhost ~]# yum list Oraclelinux-developer-release-el7
```

Here is an example:

```
[root@zdmhost ~]# yum list expect
Loaded plugins: langpacks, ulninfo
Available Packages
expect.x86_64              5.45-14.
el7_1                           ol7_latest

[root@zdmhost ~]# yum install -y expect.x86_64
```

Creating a zdm group and users

Create an OS group named `zdm` and a user named `zdmuser`:

```
[root@zdmhost ~]# whoami
 root
```

```
[root@zdmhost ~]# groupadd zdm
[root@zdmhost ~]# useradd -g zdm zdmuser
[root@zdmhost ~]# su - zdmuser
[zdmuser@zdmhost ~]$
```

Downloading ZDM software

Browse and download ZDM software from the following link, stage it in the ZDM host, and unzip it:

https://www.oracle.com/database/technologies/rac/zdm-downloads.html:

```
[zdmuser@zdmhost ~]$ pwd
/home/zdmuser
[zdmuser@zdmhost ~]$ ls -ltrh
total 741M
-rwxr-xr-x. 1 zdmuser zdm 741M Feb 21 18:00 zdm21.2.zip
[zdmuser@zdmhost ~]$ unzip zdm21.2.zip
[zdmuser@zdmhost ~]$ cd zdm21.2/
```

Creating and exporting directories

ZDM software installation requires the creation of certain directories and exporting them as environment variables. We need to create ZDM_BASE and ZDM_HOME, where ZDM_HOME is the path to the binary installation. You can also add these environment variables to the zdmuser login profile.

Installing ZDM

This is pretty straightforward, as shown in the following instruction set:

```
[zdmuser@zdmhost zdm21.2]$ mkdir /home/zdmuser/zdmbase /home/
zdmuser/zdmhome
[zdmuser@zdmhost zdm21.2]$ export ZDM_BASE=/home/zdmuser/
zdmbase
[zdmuser@zdmhost zdm21.2]$ export ZDM_HOME=/home/zdmuser/
zdmhome
[zdmuser@zdmhost zdm21.2]$ ./zdminstall.sh setup
oraclehome=$ZDM_HOME oraclebase=$ZDM_BASE ziploc=/home/zdmuser/
zdm21.2/zdm_home.zip -zdm
```

This will install the zdm binary into the ZDM_HOME path and create an OS service called zdmservice, which by default is not started.

```
[zdmuser@zdmhost zdm21.2]$ cd $ZDM_HOME/bin
[zdmuser@zdmhost bin]$ ./zdmservice status


-------------------------------------------
        Service Status
-------------------------------------------

  Running:       false
  Tranferport:
  Conn String:   jdbc:mysql://localhost:8897/
  RMI port:      8895
  HTTP port:     8896
  Wallet path:   /home/zdmuser/zdmbase/crsdata/zdmhost/security

[zdmuser@zdmhost bin]$
```

Figure 3.5: Service status

We need to manually start the service using the zdmservice CLI start option and check the status:

```
[zdmuser@zdmhost zdm21.2]$ cd $ZDM_HOME/bin
[zdmuser@zdmhost bin]$ ./zdmservice start
```

Note that ZDM internally uses MySQL, and as part of the installation step, it configures the service to the default port of 8897. You can check the port number by running the status command.

Target Autonomous Database creation

Since the target database we are planning to migrate our source database to is an autonomous transaction-processing database in OCI, we should therefore complete the provision of the autonomous database. That includes creating OCI users that would have privileges to access and create groups, compartments, and a VCN, defining IAM policies, and then provisioning the Autonomous Database. You may have to configure additional network configuration between your on-premises data center and Autonomous Database, as you might prefer placing your autonomous database in a private subnet. Such discussions have been omitted from this chapter, as they are beyond its scope. It is expected that you have a prior understanding of OCI and its various component services before you plan to work on actual database migration to Autonomous Database.

Generating an OCI API key pair in the pem format

We need to create and use an OCI API key pair to authenticate access to OCI services when invoking from the OCI CLI/SDKs. On the Linux platform, you can follow the following steps to create an OCI API RSA key pair in the pem format without a passphrase. For Windows, you can use Git Bash and run the commands, for which you must first install Git Bash for Windows. Here, we demonstrate the steps to create an RSA key pair in the pem format in zdmhost that will be used by the ZDM service. Follow the following instructions to create the key pair in the pem format without a passphrase:

```
# Create .oci directory in zdmuser home directory
[zdmuser@zdmhost ~]$ mkdir ~zdmuser/.oci
[zdmuser@zdmhost ~]$ cd ~zdmuser/.oci


# create private key in pem format without passphrase

[zdmuser@zdmhost .oci]$ openssl genrsa -out ~/.oci/oci_api_key.
pem 2048
[zdmuser@zdmhost .oci]$ chmod 600 oci_api_key.pem


# create public key in pem format from the private key file

[zdmuser@zdmhost .oci]$ openssl rsa -pubout -in oci_api_key.pem
-out oci_api_key_public.pem


# View the content of public pem file.
[zdmuser@zdmhost .oci]$ cat oci_api_key_public.pem


# Copy output to be used when adding public key in OCI for
authenticating API request
```

You need to copy the output of the cat command for the public key and add it to the OCI logged-in user API keys by clicking on the **Add API Key** button, as shown in the following figure:

Figure 3.6: Add API Key

This will open up a popup to provide the public key file in the pem format. It gives you three options, as shown in the following figure:

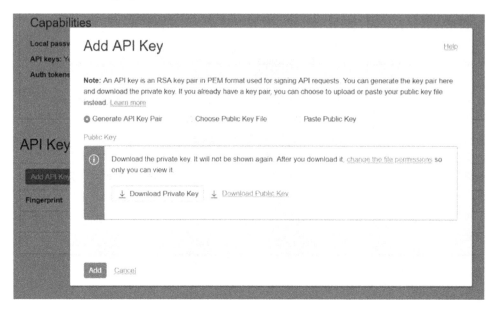

Figure 3.7: Adding a public API

Since we already have the file saved locally on our desktop, we will choose the second option and select the public key file in the pem format, as shown in the following figure:

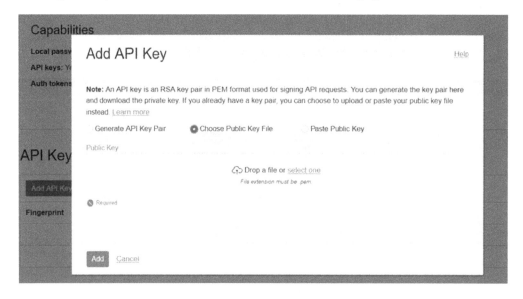

Figure 3.8: Adding a public API

As you can see in the figure, the `publicKey.pem` file has been selected:

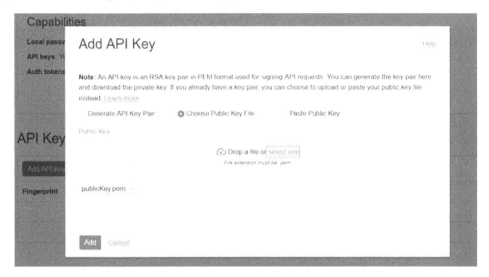

Figure 3.9: publicKey.pem

Click **Add**, which opens up a pop-up confirmation screen, as shown in *Figure 3.10*. Copy the contents under the **Configuration File Preview Read-Only** box, as they will be used to create the CLI config file in the `zdm` host.

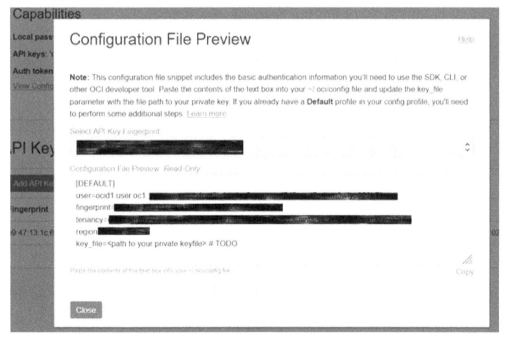

Figure 3.10: AddPublicAPI_Key_5

Click **Close**. This will store the file with fingerprint information, as shown in *Figure 3.11*. Save this fingerprint information to be used later while preparing the response file.

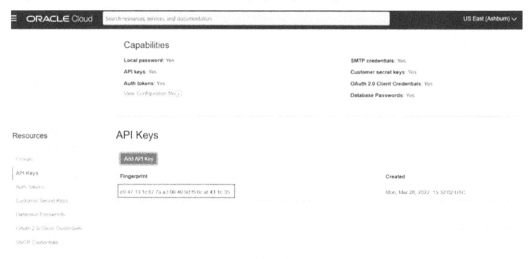

Figure 3.11: AddPublicAPI_Key_6

As the concluding step of OCI API key pair creation, you must create the `config` file that will be used by the OCI CLI. Using the copied contents, as seen in *Figure 3.11*, create a config file in the `.oci` directory and fill in the respective values from your OCI-privileged user account:

```
# Create .oci directory in zdmuser home directory
[zdmuser@zdmhost ~]$ cd ~zdmuser/.oci
[zdmuser@zdmhost ~]$ vi config

user=<fill it>
fingerprint=<fill it>
tenancy_id=<fill it>
region=<fill it>
key_file=/home/zdmuser/.oci/oci_api_key.pem
```

Fill in the values and save the config file.

Installing the OCI CLI

These steps can be directly referred to in the OCI documentation. For the sake of simplicity, the steps performed to install the OCI CLI are mentioned as follows. These steps should be performed either as the root user or, if `zdmuser` has `sudo` privilege, they can be directly executed, as shown here:

```
# Create .oci directory in zdmuser home directory

[zdmuser@zdmhost ~]$ sudo yum install -y python36-oci-cli
[zdmuser@zdmhost ~]$ oci -version

# Test command line

[zdmuser@zdmhost ~]$ oci iam region list
```

Preparing an SSH key pair

In order to freely log in from `zdmhost` to a source database server or vice versa, generate a passwordless SSH key pair and use the same key pair for both the source database server and `zdm` host. You can find a lot of online references on how to create an SSH key pair. For the sake of simplicity, store the same public key in both the source database server and the `zdm` host `.SSH` directory. Validate passwordless access from `zdmhost` to the source database server and vice versa.

The following represents an example of how `zdmuser` logs in from `zdmhost` to an on-premises database host:

```
[zdmuser@zdmhost ~]$ ssh -i /home/zdmuser/.ssh/privateKey opc@
opdb
```

Creating an Object Store bucket

During the migration process, as part of the initial synchronization of the source database to the target database, the migration process creates a Data Pump backup of the database schemas and stores it in an OCI Object Store bucket. Here, we are not going to see the entire steps for how to create an OCI Standard Object Store bucket, but instead just a two-point reference to how it looks after creating the storage bucket.

As shown in *Figure 3.12*, navigate to **Buckets** in the OCI console:

Figure 3.12: OCI_ObjectStore_Bucket_1

This will list the previously created bucket, `zdm_bucket`, as shown in *Figure 3.13*:

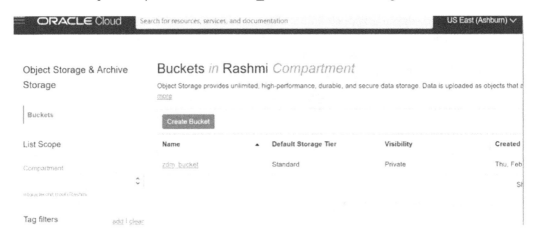

Figure 3.13: OCI_ObjectStore_Bucket_2

Creating an OCI user authentication token

Data dumps stored in an Object Storage bucket can be uploaded into Autonomous Database. This is done using the Autonomous Database APIs that can authenticate with the OCI Object Store bucket, read the files, and upload them to Autonomous Database. This authentication mechanism is achieved by creating an authentication token for the OCI user. See the following figure:

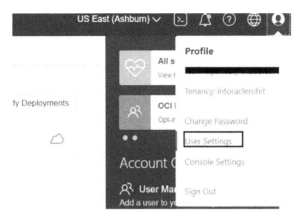

Figure 3.14: Authentication_Token_Navigate

This generates the following token:

Figure 3.15: Authentication_Token_Generate

Such a generated token looks like the following:

Figure 3.16: Authentication_Token_Generated

Keeping the same time zone

You must validate that the source database, ZDM service host, and target database are all in the same time zone. If not, liaise with your system administrator to ensure they are all in the same time zone before attempting to perform the migration.

Database character set

You must ensure that the database character set of the source and target are the same for the migration to be successful.

Setting up the GoldenGate Microservices hub in the OCI marketplace

Let's go through each step in detail.

Provisioning

Zero-downtime logical migration to Autonomous Database can be performed using Oracle GoldenGate. In this migration process, we will provision Oracle GoldenGate microservices from the OCI Marketplace. Completing the provisioning of the Autonomous Database prior to provisioning Oracle GoldenGate microservices will automatically import the Autonomous Database wallet, while setting the environment for deployment. Here, we will learn about the basic steps to provision Oracle GoldenGate in OCI and perform the required configuration. I would also advise you to refer to the public-facing document for Oracle GoldenGate to get a more detailed understanding of its capabilities.

Let's now move on to learn about the steps involved in provisioning Oracle GoldenGate. Log in to your OCI account and navigate to **All Applications** under **Marketplace**, as shown in the following figure:

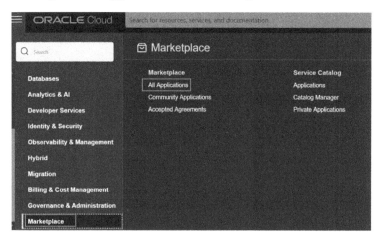

Figure 3.17: OCI_Marketplace

In the next window, search for `goldengate`, as shown in the following figure, which will shortlist applications with that naming pattern. Choose the application listed as **Oracle GoldenGate for Oracle – Database Migrations**, which has a price mentioned as **Free**. This service can be used for 183 days without charge; hence, you should prepare a proper migration plan for all the databases that are going to be migrated using ZDM, in the logical migration process. If your current **Oracle Golden Gate for Oracle – Database Migrations** service exceeds the mandated 183 days, then the service instance will become inactive, in which case you have to provision another instance of the service and redo all the configurations you made earlier for the purpose of migration; hence, it is very important to have a proper migration strategy to migrate databases within the service timeline limit. Note that it's the service that is free but not the compute on which the service will be provisioned; hence, you must pay for the compute and storage. The prerequisite is to have a VCN provisioned before you provision golden gate instance. And you must have the required compute resources where your Oracle Golden Gate service would be deployed.

Figure 3.18: Search_GoldenGate

You can find the software price per OCPU, as shown here:

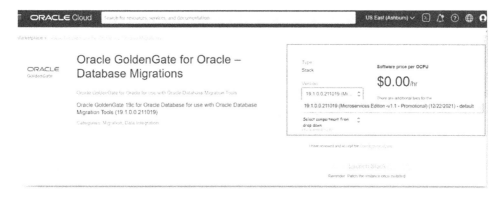

Figure 3.19: GoldenGate_Microservices_For_DB_Migration

As highlighted in the box in *Figure 3.19*, the charge you pay for this service is \$0. The current version of Oracle GoldenGate Microservices available is 19.1.0.0.211019. From the dropdown, select the compartment where you already have your Autonomous Database instance in order to automate some of the prerequisite actions.

Check the checkbox that says **I have reviewed and accept the Oracle terms of use**, and this will enable **Launch Stack**, which you can click on, as shown in the following figure:

Figure 3.20: Launch_OGGMA_Stack

Fill in the details for the stack, as shown here:

Figure 3.21: Fill_OGGMA_Stack_1

Click **Next** and fill in the required details, as per your environment. As shown in *Figure 3.22*, fill in **Network Settings**:

Figure 3.22: Fill_OGGMA_Stack_Nw

The following figure shows the chosen availability domain and shape based on your available quota limits:

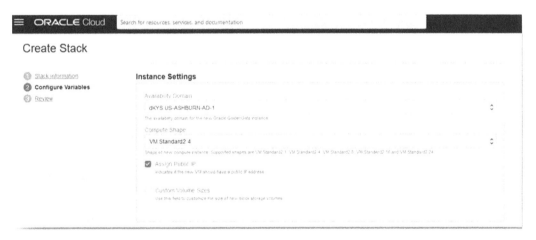

Figure 3.23: Fill_OGGMA_Stack_Inst

The next figure is a crucial one where you choose the deployment – that is, the source is your on-premises database and the target is your Autonomous Database deployment:

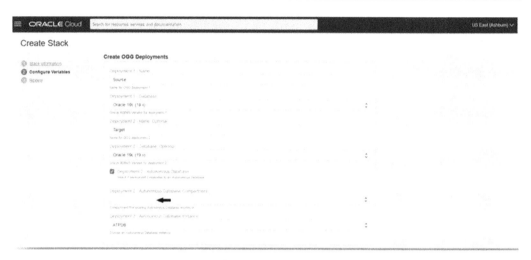

Figure 3.24: Fill_OGGMA_Stack_Deployment

Fill in the SSH public key and check the summary of your inputs using which the stack is ready for execution using Terraform, as shown in the following figure:

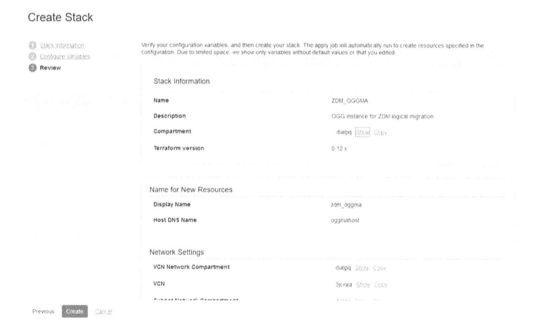

Figure 3.25: OGGMA_Stack_Summary_1

Here's another example of the stack summary:

Figure 3.26: OGGMA_Stack_Summary_2

Click **Create**. This will submit a resource manager job and provision the necessary resources, using the inputs you provided for the instance creation. You can wait for its completion, as shown in *Figure 3.26*:

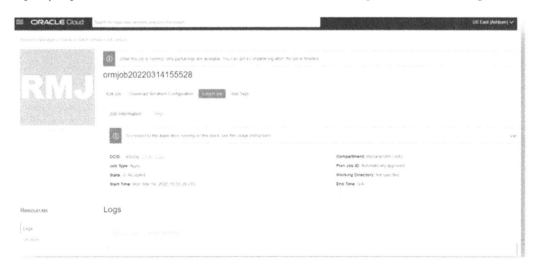

Figure 3.27: OGGMA_Stack_RMJ

Meanwhile, you can monitor the logs to get the IP of the OGG microservices compute, as shown in *Figure 3.28*:

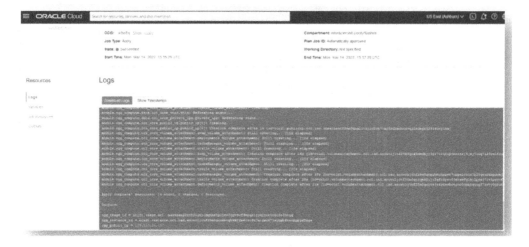

Figure 3.28: OGGMA_Stack_RMJ_Logs

After completion of the job, navigate to the compute instance section in OCI, where you can see the compute instance provisioned for Oracle GoldenGate, as shown in *Figure 3.29*:

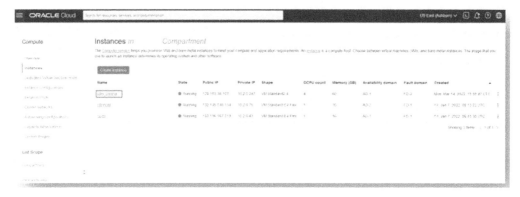

Figure 3.29: OGGMA_Compute

Click on the compute to view the details, as shown in *Figure 3.30*:

Figure 3.30: OGGMA_Compute_Details

Copy the public IP and log in to the host as the OPC user using the SSH private key pair of the public key that you uploaded during the provisioning steps. Run the OS "cat" command to check the contents of the /home/opc/ ogg-credentials.json file. This file contains the initial login credentials to log in to the OGGMA UI in the browser, as shown in *Figure 3.30*:

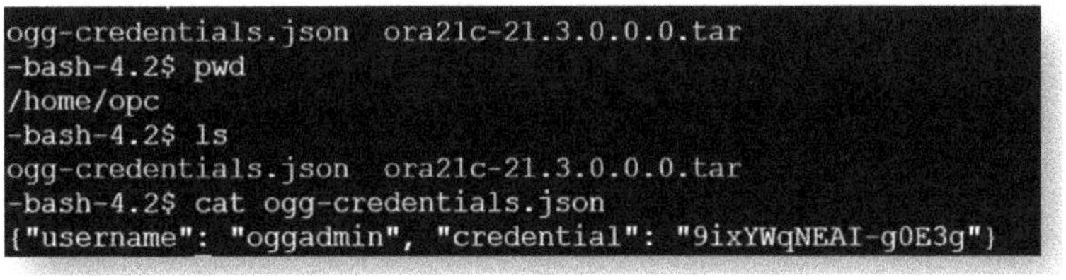

```
ogg-credentials.json   ora21c-21.3.0.0.0.tar
-bash-4.2$ pwd
/home/opc
-bash-4.2$ ls
ogg-credentials.json   ora21c-21.3.0.0.0.tar
-bash-4.2$ cat ogg-credentials.json
{"username": "oggadmin", "credential": "9ixYWqNEAI-g0E3g"}
```

Figure 3.31: oggadmin_Initial_Credential_.jpg

You will initially use these credentials for first-time login to the OGGMA Service Manager or the admin server after logging in. You can change the password after you log in.

Open the Service Manager using the public IP in the URL as shown in the compute – for example, https://129.153.36.107/. Accept the risk and continue.

To log in, use oggadmin as the username and the same password contained in the ogg-credentials. json file, as shown here:

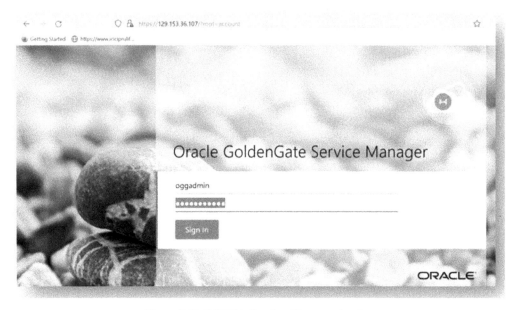

Figure 3.32: OGGMA_Service_Manager_Login

This will show the different deployments and associated ports for each deployment, as shown here:

Figure 3.33: OGGMA_Service_Manager

The **Change Password** settings can be found in the path, as shown here:

Figure 3.34: OGGMA_Service_Manager_ChangePwd

You can change the password by clicking on **Change Password**, as shown in the top right corner of the previous figure. Similarly, you can change the password for the `oggadmin` user after you log in to the administration server for source deployment (click on port `9011`) and target deployment (click on port `9021`). After changing the password, you will have to log in again with a new password.

GoldenGate configuration in the source database

After provisioning, the next step requires configuring the database setting for Oracle GoldenGate before ZDM can use it to prepare the zero-downtime logical replication.

Once patching is done, the following are a few of the database configuration steps required for Oracle GoldenGate, and if you are familiar with it, then you should already be familiar with the complete steps.

Execute the following steps in the source database, as shown in the following code block:

```
SQL> sqlplus / as sysdba


# If database is not in archivelog mode
SQL> shutdown immediate
SQL> startup mount
SQL> alter database archivelog;


# enable force logging and supplemental logging
SQL> alter database force logging;
SQL> alter database add supplemental log data;
SQL> alter system set enable_goldengate_replication=true;
SQL> alter system set streams_pool_size=2048M;  # for running
integrated extract
```

```
SQL> alter database open;
SQL> archive log list
SQL> sho parameter goldengate
SQL> alter system switch logfile;
SQL> select supplemental_log_data_min, force_logging,log_mode
from v$database;


# To process certain updates for object types like user-defined
types, Nested tables, XML Types #objects, Oracle Golden Gate
uses flashback queries on UNDO tablespaces, hence you may set
#below if you have these data types


SQL> alter system set undo_management=auto;
SQL> alter system set undo_retention=86400; # set the value as
suitable


# For sizing of undo tablespaces and LOB data types, please
check the golden gate documentation under section "Preparing
the database for Oracle GoldenGate" for guidance on additional
configuration steps
```

Next, create an Oracle GoldenGate admin user in the source database that will be responsible for capturing real-time database changes. If you have a container database, then you must have two sets of users, as explained here:

```
***** Source Database User credentials for Oracle GoldenGate
configuration ******
# If source database is a multitenant database, you have to
create a goldengate administrator user in the container as well
as pluggable database. In case of non-container database, just
create a goldengate administrator user. After that, you must
grant specific privileges to that user
# In container database
sqlplus / as sysdba
# Prefix "c##" in user name, as in this example
SQL> create user c##ggadmin identified by <&pwd> default
tablespace users temporary tablespace temp;
SQL> grant connect, resource to c##ggadmin;
SQL> grant unlimited tablespace to c##ggadmin;
```

```
SQL> alter user c##ggadmin quota unlimited ON USERS;
SQL> grant select any dictionary to c##ggadmin;
SQL> grant create view to c##ggadmin;
SQL> grant execute on dbms_lock to c##ggadmin;
SQL> exec dbms_goldengate_auth.grant_admin_
privilege('C##GGADMIN',container=>'ALL');


# In pluggable database or if it's a non-container database
(login as sys user)
SQL> alter session set container=&pdb ; --applicable for
pluggable database only
SQL> create user ggadmin identified by <&pwd> default
tablespace users temporary tablespace temp;
SQL> grant connect, resource to ggadmin;
SQL> alter user ggadmin quota 100M ON USERS;
SQL> grant unlimited tablespace to ggadmin;
SQL> grant select any dictionary to ggadmin;
SQL> grant create view to ggadmin;
SQL> grant execute on dbms_lock to ggadmin;
SQL> exec dbms_goldengate_auth.GRANT_ADMIN_
PRIVILEGE('GGADMIN');
```

When you have completed the user creation, you have to install the UTL_SPADV or UTL_RPADV (in 19c) package in the source database (a container database in the case of a multitenant) for Integrated Extract performance analysis:

```
# Command to be executed taking the multitenant database into
example.
[oracle@opdb ~]$ cd $ORACLE_HOME/rdbms/admin
[oracle@opdb ~]$ sqlplus / as sysdba
SQL> grant execute on DBMS_LOGREP_UTIL to C##GGADMIN;
SQL> grant select on SYSTEM.AQ$_QUEUE_TABLES to C##GGADMIN;
SQL> grant select on SYSTEM.AQ$_QUEUES to C##GGADMIN;
SQL> grant execute on DBMS_LOGREP_UTIL to C##GGADMIN;
SQL> conn c##ggadmin@<pdb>
SQL> @utlrpadv.sql
```

So, this completes all the steps that are required for source configuration.

Preparing a target deployment DB environment

In this scenario, the target database chosen for migration is Oracle Autonomous Database. Connection to Autonomous Database is protected using a wallet. In order to establish a connection to the Autonomous Database from the Oracle GoldenGate target deployment, you have to download your Autonomous Database wallet file from the Autonomous Database details page in the OCI console and transfer it to the `/u02/deployments/Target/etc/adb` location in the GoldenGate compute, and extract the ZIP file to the same location, as shown in the following figure:

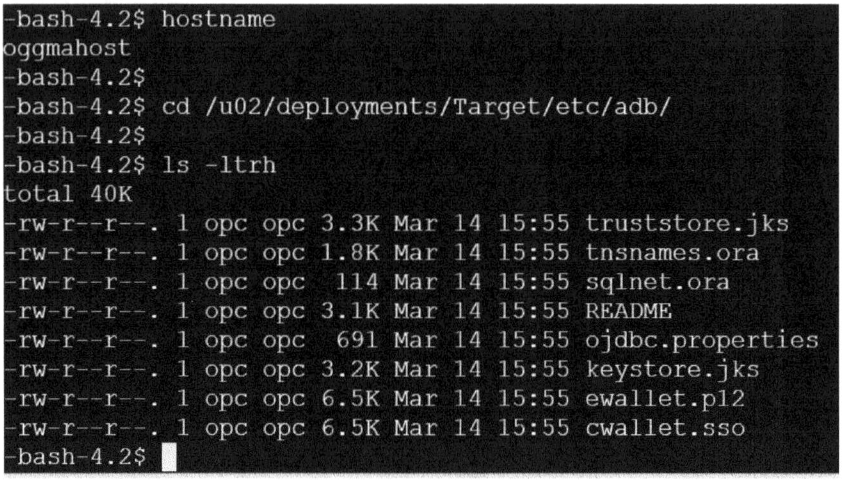

```
-bash-4.2$ hostname
oggmahost
-bash-4.2$
-bash-4.2$ cd /u02/deployments/Target/etc/adb/
-bash-4.2$
-bash-4.2$ ls -ltrh
total 40K
-rw-r--r--. 1 opc opc 3.3K Mar 14 15:55 truststore.jks
-rw-r--r--. 1 opc opc 1.8K Mar 14 15:55 tnsnames.ora
-rw-r--r--. 1 opc opc  114 Mar 14 15:55 sqlnet.ora
-rw-r--r--. 1 opc opc 3.1K Mar 14 15:55 README
-rw-r--r--. 1 opc opc  691 Mar 14 15:55 ojdbc.properties
-rw-r--r--. 1 opc opc 3.2K Mar 14 15:55 keystore.jks
-rw-r--r--. 1 opc opc 6.5K Mar 14 15:55 ewallet.p12
-rw-r--r--. 1 opc opc 6.5K Mar 14 15:55 cwallet.sso
-bash-4.2$
```

Figure 3.35 – GoldenGate compute

During migration, this wallet file will be used to set up a database connection in Oracle GoldenGate.

Activating a target GoldenGate Admin user

Oracle Autonomous Database by default includes the GoldenGate admin user with the name ggadmin, which has a locked account status. You have to unlock the user account with an admin user login. By default, one of the GoldenGate database parameters named enable_goldengate_replication is also enabled. Hence, you can also check the status of the parameter by querying the database. The following are the steps that you can follow to unlock and check the parameter value:

```
Connect with admin user in autonomous database and unlock the
ggadmin account
sqlplus admin@<atp db name>_low
SQL> select username,account_status from dba_users where
username='GGADMIN';
SQL> alter user ggadmin identified by Oracle123456 account
unlock;
```

```
# Check the status of parameter enable_goldengate_replication
SQL> select name, value from v$parameter where name = 'enable_
goldengate_replication';
```

Setting up the OCI CLI in the source database server for file transfer

You need to set up the OCI CLI in the source database server, which will be used to transfer files from a local system to OCI object storage.

Refer to https://docs.Oracle.com/en-us/iaas/Content/API/SDKDocs/cliinstall. htm for information on setting up.

Here is a quick outline of the steps:

```
# Check OS platform release
[oracle@opdb ~]$ cat /etc/oracle-release
Oracle Linux Server release 7.9

# Install oci cli
[oracle@opdb ~]$ sudo yum install -y python36-oci-cli

#Check oci cli version
[oracle@opdb ~]$ oci --version
3.6.2

# Setup oci cli config and provide inputs, for private key,
you must provide the private key in .pem format as we created
earlier in the step.
[oracle@opdb ~]$ oci setup config
```

Preparing a ZDM response file for migration

zdm requires inputs for some parameters that it will use during migration. You can collate all these parameter file values into a response file, the template for which you can find in the ZDM_HOME directory.

Follow the following sequence of steps to create the response file:

```
[zdmuser@zdmhost ~]$ cd $ZDM_HOME/rhp/zdm/template/
[zdmuser@zdmhost template]$ cp zdm_logical_template.rsp zdm_
```

```
logical_template.rsp.orig
[zdmuser@zdmhost template]$ vi zdm_logical_template.rsp
# Set the parameter values as you can see in below example.
MIGRATION_METHOD=ONLINE_LOGICAL
DATA_TRANSFER_MEDIUM=OSS
TARGETDATABASE_OCID=<OCID of target database service in OCI>
# Fill the parameter values under each group
#General inputs (Source and target database details, connection
details admin user details etc)
#GoldenGate parameter settings
#Datapump settings
```

A dry run of ZDM

In this step, you do a dry run of the migration process using ZDM:

```
# In ZDM service host
[zdmuser@zdmhost ~]$cd $ZDM_HOME/bin

[zdmuser@zdmhost bin]$ ./zdmcli migrate database -rsp /home/
zdmuser/zdmhome/rhp/zdm/template/zdm_logical_online.rsp \
> -sourcesid orcl \
> -sourcenode xxx.xxx.xxx.xxx \
> -srcauth zdmauth \
> -srcarg1 user:opc \
> -srcarg2 identity_file:/home/zdmuser/.ssh/privateKey \
> -srcarg3 sudo_location:/bin/sudo \
> -eval -skipadvisor
```

This will create an audit ID for the execution and prompts you for a password for the different database users, such as the system, ggadmin, oggadmin, and the OCI user OCID, and then create a job ID for the execution, as shown in *Figure 3.36*:

```
[zdmuser@zdmhost bin]$ ./zdmcli migrate database -rsp /home/zdmuser/zdmhome/rhp/zdm/template/zdm_logical_online.rsp \
> -sourcesid orcl \
> -sourcenode                  \
> -srcauth zdmauth \
> -srcarg1 user:opc \
> -srcarg2 identity_file:/home/zdmuser/.ssh/privateKey \
> -srcarg3 sudo location:/bin/sudo \
> -eval -skipadvisor
zdmhost.                            Audit ID: 2
Enter source database administrative user "system" password:
Enter source database administrative user "ggadmin" password:
Enter source container database administrative user "system" password:
Enter source container database administrative user "c##ggadmin" password:
Enter target database administrative user "admin" password:
Enter target database administrative user "ggadmin" password:
Enter Oracle GoldenGate hub administrative user "oggadmin" password:
Enter Authentication Token for OCI user "ocid1.                                         ":
Operation "zdmcli migrate database" scheduled with the job ID "1".
[zdmuser@zdmhost bin]$
```

Figure 3.36: Creating an audit ID

You can check the status of the job at intervals by querying the job ID. You should see the **COMPLETED** status for each step, as shown in *Figure 3.37*:

```
./zdmcli query job -jobid <job#>
job# is the job id that was created when you invoked the
migration job.
```

```
[zdmuser@zdmhost bin]$ ./zdmcli query job -jobid 1
zdmhost.                            : Audit ID: 9
Job ID: 1
User: zdmuser
Client: zdmhost
Job Type: "EVAL"
Scheduled job command: "zdmcli migrate database -rsp /home/zdmuser/zdmhome/rhp/zdm/template/zdm_logical_online.rsp -sourcesid orcl -sou
rcenode                srcauth zdmauth -srcarg1 user:opc -srcarg2 identity_file:/home/zdmuser/.ssh/privateKey -srcarg3 sudo_location:
/bin/sudo -eval -skipadvisor"
Scheduled job execution start time: 2022-04-12T18:10:48Z. Equivalent local time: 2022-04-12 18:10:48
Current status: SUCCEEDED
Result file path: "/home/zdmuser/zdmbase/chkbase/scheduled/job-1-2022-04-12-18:11:14.log"
Job execution start time: 2022-04-12 18:11:14
Job execution end time: 2022-04-12 18:52:47
Job execution elapsed time: 1 minutes 16 seconds
ZDM_VALIDATE_SRC ..................... COMPLETED
ZDM_VALIDATE_TGT ..................... COMPLETED
ZDM_VALIDATE_GG_HUB .................. COMPLETED
ZDM_VALIDATE_DATAPUMP_SETTINGS_SRC .... COMPLETED
ZDM_VALIDATE_DATAPUMP_SETTINGS_TGT .... COMPLETED
ZDM_PREPARE_DATAPUMP_SRC ............. COMPLETED
ZDM_DATAPUMP_ESTIMATE_SRC ............ COMPLETED
[zdmuser@zdmhost bin]$
```

Figure 3.37: zdm_eval_job_status

If any of the steps resulted in a failed status, you can work on to fix the issue and rerun the dry run until you see each step has a **COMPLETED** status.

Run the migration job

Once you see the status has completed in the dry run, you can proceed to run the actual database migration with ZDM, using the same command without the -eval option:

```
# In ZDM service host
[zdmuser@zdmhost ~]$cd $ZDM_HOME/bin
```

```
[zdmuser@zdmhost bin]$ ./zdmcli migrate database -rsp /home/
zdmuser/zdmhome/rhp/zdm/template/zdm_logical_online.rsp \
> -sourcesid orcl \
> -sourcenode xxx.xxx.xxx.xxx \
> -srcauth zdmauth \
> -srcarg1 user:opc \
> -srcarg2 identity_file:/home/zdmuser/.ssh/privateKey \
> -srcarg3 sudo_location:/bin/sudo \
> -skipadvisor
```

Similarly, it will prompt you for inputs and then submit a job, as shown in *Figure 3.38*:

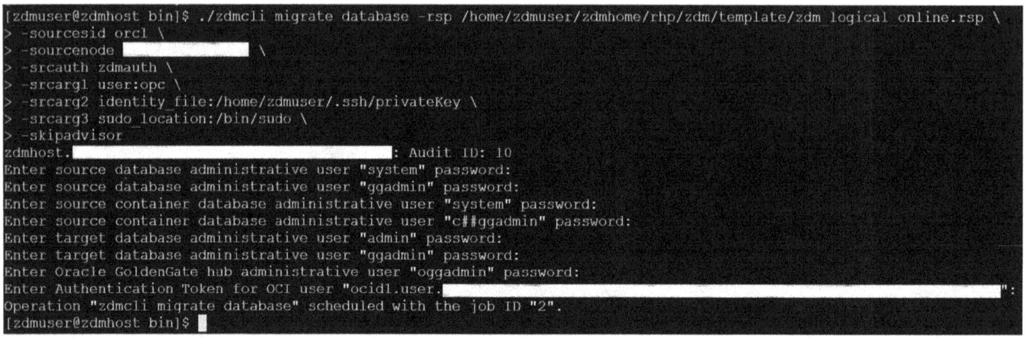

Figure 3.38: Database_migration_using_zdmcli

This will create your GoldenGate source and target credentials, create GoldenGate extracts, run an initial load, and then run replication into the target autonomous database, allowing you to switch your application connection to the new environment and then clean up the environment.

In the following figure, note that there was an error creating the GoldenGate capture process due to an "insufficient privilege" error. You can log in to the source and run the database commands to grant the appropriate privilege to the ggadmin user. Once that issue is fixed, you can invoke zdmcli once more with the same parameters, and it will again create a new job and start executing it.

Figure 3.39: Golden_Gate_extract.jpg

You can go back to the terminal and query `job_id`, where you can see the series of steps it is executing and the status of the execution at each step. You can see that the step names are quite self-explanatory – for example, `ZDM_ADD_HEARTBEAT_SRC`, `ZDM_PREPARE_GG_HUB`, and `ZDM_DATAPUMP_EXPORT_SRC`, which indicate the creation of a heartbeat table in Oracle GoldenGate, preparing the GoldenGate Hub, and taking `expdp` from the source database respectively, and it also indicates whether the step was completed successfully or failed. It has different statuses such as **COMPLETED**, **STARTED**, and **PENDING**, as you can see in the following figure:

```
[zdmuser@zdmhost bin]$ ./zdmcli query job -jobid 2
zdmhost.                                : Audit ID: 13
Job ID: 2
User: zdmuser
Client: zdmhost
Job Type: "MIGRATE"
Scheduled job command: "zdmcli migrate database -rsp /home/zdmuser/zdmhome/rhp/zdm/template/zdm_logical_online.rsp -sourcesid orcl -sou
rcenode                      -srcauth zdmauth -srcarg1 user:opc -srcarg2 identity_file:/home/zdmuser/.ssh/privateKey -srcarg3 sudo_location:
/bin/sudo -skipadvisor"
Scheduled job execution start time: 2022-04-13T06:18:13Z. Equivalent local time: 2022-04-13 06:18:13
Current status: EXECUTING
Current Phase: "ZDM_CREATE_GG_EXTRACT_SRC"
Result file path: "/home/zdmuser/zdmbase/chkbase/scheduled/job-2-2022-04-13-06:18:16.log"
Job execution start time: 2022-04-13 06:18:16
Job execution end time: 2022-04-13 06:21:33
Job execution elapsed time: 3 minutes 16 seconds
ZDM_VALIDATE_SRC .................... COMPLETED
ZDM_VALIDATE_TGT .................... COMPLETED
ZDM_VALIDATE_GG_HUB ................. COMPLETED
ZDM_VALIDATE_DATAPUMP_SETTINGS_SRC .... COMPLETED
ZDM_VALIDATE_DATAPUMP_SETTINGS_TGT ... COMPLETED
ZDM_PREPARE_DATAPUMP_SRC ............ COMPLETED
ZDM_DATAPUMP_ESTIMATE_SRC ........... COMPLETED
ZDM_PREPARE_GG_HUB .................. COMPLETED
ZDM_ADD_HEARTBEAT_SRC ............... COMPLETED
ZDM_ADD_SCHEMA_TRANDATA_SRC ......... COMPLETED
ZDM_CREATE_GG_EXTRACT_SRC ........... STARTED
ZDM_PREPARE_DATAPUMP_TGT ............ PENDING
ZDM_DATAPUMP_EXPORT_SRC ............. PENDING
ZDM_UPLOAD_DUMPS_SRC ................ PENDING
ZDM_DATAPUMP_IMPORT_TGT ............. PENDING
ZDM_POST_DATAPUMP_SRC ............... PENDING
ZDM_POST_DATAPUMP_TGT ............... PENDING
ZDM_ADD_HEARTBEAT_TGT ............... PENDING
ZDM_ADD_CHECKPOINT_TGT .............. PENDING
```

Figure 3.40: zdmcli_migration_steps

Note here that it will create a bucket to store the source dump files in object storage, as the target is Autonomous Database. In *Figure 3.41*, you can see the list of Data Pump dump files uploaded automatically by ZDM into OCI object storage:

Figure 3.41: Data_dump_in_ObjectStorageBucket

After completion of initial loading using Data Pump, it will create a target replication process, as shown in *Figure 3.42*:

```
[zdmuser@zdmhost bin]$ ./zdmcli query job -jobid 2
zdmhost:                                          : Audit ID: 14
Job ID: 2
User: zdmuser
Client: zdmhost
Job Type: "MIGRATE"
Scheduled job command: "zdmcli migrate database -rsp /home/zdmuser/zdmhome/rhp/zdm/template/zdm_logical_online.rsp -sourcesid orcl -sou
rcenode              -srcauth zdmauth -srcarg1 user:opc -srcarg2 identity_file:/home/zdmuser/.ssh/privateKey -srcarg3 sudo_location:
/bin/sudo -skipadvisor"
Scheduled job execution start time: 2022-04-13T06:18:13Z. Equivalent local time: 2022-04-13 06:18:13
Current status: EXECUTING
Current Phase: "ZDM_CREATE_GG_REPLICAT_TGT"
Result file path: "/home/zdmuser/zdmbase/chkbase/scheduled/job-2-2022-04-13-06:18:16.log"
Job execution start time: 2022-04-13 06:18:16
Job execution end time: 2022-04-13 06:21:33
Job execution elapsed time: 3 minutes 16 seconds
ZDM_VALIDATE_SRC ..................... COMPLETED
ZDM_VALIDATE_TGT ..................... COMPLETED
ZDM_VALIDATE_GG_HUB .................. COMPLETED
ZDM_VALIDATE_DATAPUMP_SETTINGS_SRC ... COMPLETED
ZDM_VALIDATE_DATAPUMP_SETTINGS_TGT ... COMPLETED
ZDM_PREPARE_DATAPUMP_SRC ............. COMPLETED
ZDM_DATAPUMP_ESTIMATE_SRC ............ COMPLETED
ZDM_PREPARE_GG_HUB ................... COMPLETED
ZDM_ADD_HEARTBEAT_SRC ................ COMPLETED
ZDM_ADD_SCHEMA_TRANDATA_SRC .......... COMPLETED
ZDM_CREATE_GG_EXTRACT_SRC ............ COMPLETED
ZDM_PREPARE_DATAPUMP_TGT ............. COMPLETED
ZDM_DATAPUMP_EXPORT_SRC .............. COMPLETED
ZDM_UPLOAD_DUMPS_SRC ................. COMPLETED
ZDM_DATAPUMP_IMPORT_TGT .............. COMPLETED
ZDM_POST_DATAPUMP_SRC ................ COMPLETED
ZDM_POST_DATAPUMP_TGT ................ COMPLETED
ZDM_ADD_HEARTBEAT_TGT ................ COMPLETED
ZDM_ADD_CHECKPOINT_TGT ............... COMPLETED
ZDM_CREATE_GG_REPLICAT_TGT ........... STARTED
```

Figure 3.42: add_OGG_replicat

In *Figure 3.43*, you can see that the replication process is created in the target deployment (port 9021 on the Service Manager screen). This process will run continuously until all the replication is done and it is in real-time sync with the source database.

Figure 3.43: add_replicat

Once all the steps have been completed, they will each show a **COMPLETED** status, as shown in *Figure 3.44*:

```
[zdmuser@zdmhost bin]$ ./zdmcli query job -jobid 2
zdmhost                         : Audit ID: 16
Job ID: 2
User: zdmuser
Client: zdmhost
Job Type: "MIGRATE"
Scheduled job command: "zdmcli migrate database -rsp /home/zdmuser/zdmhome/rhp/zdm/template/zdm_logical_online.rsp -sourcesid orcl -sou
rcenode.          -srcauth zdmauth -srcarg1 user:opc -srcarg2 identity_file:/home/zdmuser/.ssh/privateKey -srcarg3 sudo_location:
/bin/sudo -skipadvisor"
Scheduled job execution start time: 2022-04-13T06:18:13Z. Equivalent local time: 2022-04-13 06:18:13
Current status: SUCCEEDED
Result file path: "/home/zdmuser/zdmbase/chkbase/scheduled/job-2-2022-04-13-06:18:16.log"
Job execution start time: 2022-04-13 06:18:16
Job execution end time: 2022-04-13 06:59:44
Job execution elapsed time: 25 minutes 14 seconds
ZDM_VALIDATE_SRC ..................... COMPLETED
ZDM_VALIDATE_TGT ..................... COMPLETED
ZDM_VALIDATE_GG_HUB .................. COMPLETED
ZDM_VALIDATE_DATAPUMP_SETTINGS_SRC ... COMPLETED
ZDM_VALIDATE_DATAPUMP_SETTINGS_TGT ... COMPLETED
ZDM_PREPARE_DATAPUMP_SRC ............. COMPLETED
ZDM_DATAPUMP_ESTIMATE_SRC ............ COMPLETED
ZDM_PREPARE_GG_HUB ................... COMPLETED
ZDM_ADD_HEARTBEAT_SRC ................ COMPLETED
ZDM_ADD_SCHEMA_TRANDATA_SRC .......... COMPLETED
ZDM_CREATE_GG_EXTRACT_SRC ............ COMPLETED
ZDM_PREPARE_DATAPUMP_TGT ............. COMPLETED
ZDM_DATAPUMP_EXPORT_SRC .............. COMPLETED
ZDM_UPLOAD_DUMPS_SRC ................. COMPLETED
ZDM_DATAPUMP_IMPORT_TGT .............. COMPLETED
ZDM_POST_DATAPUMP_SRC ................ COMPLETED
ZDM_POST_DATAPUMP_TGT ................ COMPLETED
ZDM_ADD_HEARTBEAT_TGT ................ COMPLETED
ZDM_ADD_CHECKPOINT_TGT ............... COMPLETED
ZDM_CREATE_GG_REPLICAT_TGT ........... COMPLETED
ZDM_MONITOR_GG_LAG ................... COMPLETED
ZDM_SWITCHOVER_APP ................... COMPLETED
ZDM_RM_GG_EXTRACT_SRC ................ COMPLETED
```

Figure 3.44: All_steps_status

The last step involves cleaning up all the created resources. You can perform a validation of the migration by querying the schema and its objects in the target autonomous database, as shown in *Figure 3.45*. Later, you can compare it with the source schema.

```
SQL> sho con_name

CON_NAME
--------------------------------
AAVQREAGYXPPPFY_ATPDB
SQL>
SQL> select object_type, count(*) from dba_objects where owner='HR' group by object_type;

OBJECT_TYPE                   COUNT(*)
----------------------------- ----------
INDEX                               19
TRIGGER                              2
PROCEDURE                            2
SEQUENCE                             3
TABLE                                7
VIEW                                 1

6 rows selected.
```

Figure 3.45: Schema_objects_in_target_db

This results in us successfully migrating our source database into the target autonomous database with automated migration, using ZDM.

Summary

Throughout this chapter, we covered a very key topic of Autonomous Database – that is, migration to Autonomous Database if we need to leverage the rich features available within the service. We discussed some of the regular considerations when migrating a database in general, and what precautions or considerations need to be made for a successful migration. Briefly, we talked about the different migration approaches that are possible while migrating to Autonomous Database and focused on one of the recommended automated processes for migration, zdm. We split the migration into multiple parts; first, we discussed the prerequisites, followed by a detailed discussion of each of the different steps involved in this utility, and finally, we simplified the entire process by clubbing them together as one invocation, using a response file that runs the steps in sequence. A concluding part of the step was to perform a sanity check. This method allowed us to initiate multiple migrations in one go with less manual intervention. The objective of performing the individual steps is to help you understand the task accomplished in each step so that if you invoke via a response file and encounter an error, you then know what steps are involved and can easily troubleshoot and fix the issue.

OCI also provides a fully managed service, DMS, which provides a self-service experience to migrate your on-premises, Oracle Cloud, or Amazon RDS Oracle databases into co-managed or autonomous databases in OCI. Under the hood, it is entirely driven by a ZDM product that automatically handles the ZDM configurations.

In the next chapter, we will discuss how to ease out the preparation of development or a non-production database environment with refreshes of production, using the available cloud tooling.

Questions

1. What are the different methods to migrate to Autonomous Database?

2. How does zero-downtime ensure as little downtime migration as possible to Autonomous Database?

3. Where are dumps stored for import?

4. What is the minimum supported database version for migration?

5. Are you required to create new tablespaces before initiating Data Pump Import?

Answers

1. Data Pump, SQL *Loader, SQL Developer, Oracle GoldenGate, Oracle Data Integrator, and other ETL tools.

2. Using logical online migration, which automates the usage of Data Pump combined with Oracle GoldenGate.

3. An OCI Object Store bucket.

4. 11.2.0.4.

5. No, in Autonomous Database, we cannot create new tablespaces. It is restricted; we can only remap tablespaces during import. There are two tablespaces available, one for DATA and another for RECO. Besides that, in the ZDM migration method, we just have to provide the inputs in the response file, where we can also use remap_tablespace and let the ZDM tool take care of the import operation.

4

ADB Disaster Protection with Autonomous Data Guard

From our understanding of the chapters so far, **Autonomous Database** definitely is a database deployment of choice in the cloud. In this chapter, we explore more how it provides an additional level of database protection from any kind of disaster. We will also learn how to configure our Autonomous Database with a **Disaster Recovery (DR)** solution to ensure business continuity and understand the **Recovery Point Objective (RPO)** and **Recovery Time Objective (RTO)** concepts in the context of Autonomous Database. We'll also learn about the cloud tool available for automating most of the manual steps that you would otherwise do in a traditional non-cloud or **IaaS**-based deployment.

In this chapter, we will cover the following topics:

- Overview and associated terminologies
- AuDG
- Status and operations in AuDG

We will conclude this chapter by making a short summary of what we have learned in this chapter, followed by some knowledge checks to assess your understanding.

Overview and associated terminologies

To begin with, Oracle Autonomous Database has all the essential properties that significantly waive much of the data management overhead while being a cost-effective database solution. Needless to say, this is a **highly available (HA)** database solution in **Oracle Cloud Infrastructure (OCI)** with an SLA of 99.95% for a shared infrastructure and an SLO of 99.95% for Autonomous Database for a dedicated infrastructure or **Exadata Cloud at Customer (C@C)**. Thus, this makes it an ideal platform for our mission-critical application, as applications can run uninterrupted irrespective of whether there is any database maintenance activity or there is any fault in any component of the underlying infrastructure for the reasons discussed earlier.

Now, the next question that comes to mind is "Although my current Autonomous Database is HA and protected from any kind of failures, is it 100% protected?" This leads us to perform the same level of risk assessment that we would do even for an on-premises database deployment in our data centers – that is, we must have the best redundant infrastructure components, implement the best-known HA solution for our database, and all within the boundary of our data center, checking whether we are still protected from unforeseen disasters such as earthquakes, floods, and fires. The same applies to OCI as well. So, in the event of any natural disaster of this kind, our database systems should be resilient by having a DR environment, far from the primary impact site and a mirror copy of your primary database system (a standby). The answer to meet this critical requirement in Autonomous Database is fulfilled by a new capability called **Autonomous Data Guard** (**AuDG**). AuDG is a fully automated DR solution to protect your mission-critical application databases running in Oracle Autonomous Database, safeguarding them from any disaster due to site failures, human error, or data corruption.

Let's now expand our understanding of the capability in the following sections and learn about what kind of DR solution it offers and the different operational mechanisms.

AuDG

With Oracle Autonomous Database, you have the option to create a mirror copy of your primary Autonomous Database either locally or cross-region using AuDG. This is fully automated, wherein you just have to enable the option against your Autonomous Database instance and as the name sounds, under the hood, a standby copy of your primary Autonomous Database is automatically created with the needed settings for failover or switchover with zero manual intervention. This feature is available with Autonomous Database – for Shared, Dedicated, and Exadata C@C deployment options. Throughout this chapter, we will discuss this feature and gain more insight into its functionality and usage.

This chapter is dedicated to studying the feature functionality more from an Autonomous Database Shared infrastructure service perspective. The AuDG configuration for Dedicated or ExaC@C is slightly different and is not covered as part of this chapter, but if there is a need, you must refer to the OCI public documentation to understand the configuration steps:

```
https://docs.oracle.com/en/cloud/paas/autonomous-database/dedicated/
adbau/index.html#articletitle
```

The term *autonomous* is used deliberately for Data Guard and it rightfully suits the definition. By just clicking on the **Enable** link in the **Autonomous Data Guard** section of your **Autonomous Database Details** page, a standby Autonomous Database of your Autonomous Database (the primary) is created without any manual intervention. It allows the same failover and switchover capabilities as you would achieve for an on-premises database.

You can view the details of the standby instances of your primary Autonomous Database by clicking on the **Autonomous Data Guard** option under the **Resources** section, where it shows the standby instance name, its role, its state, whether available or stopped, and the region it is created in.

You have two choices to decide how you want to enable AuDG – that is, you can choose to have a local standby in the same region as your primary, a remote standby in a region (subscribed) other than your primary, or you can have both a local and cross-region standby. These types of deployment differ slightly in terms of capabilities and we will discuss this in the following section.

Local standby

In case of a local standby, AuDG creates a standby database in the same region as your primary database and the provisioning location of the standby database depends on the number of **Availability Domains (ADs)** available for the dedicated OCI region. In multi-AD OCI regions, the standby instance is created in another AD other than the AD where the primary database exists, and for single AD OCI regions, the standby database is created in another fault domain other than the fault domain where the primary database exists. This provides resilience and protection against any kind of data loss in the event that a primary database becomes unavailable. It should be noted that this is an advanced feature, only available with an Autonomous Database that is on **19c** release and above – hence if your autonomous instance is on **18c**, you need to upgrade to 19c. This is not available for Autonomous Databases provisioned with an Always Free tier subscription in OCI:

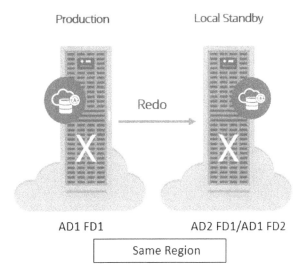

Figure 4.1: AuDG – local standby

AuDG continuously monitors the primary Autonomous Database instance and if the primary instance becomes unavailable or goes down, then AuDG automatically converts the standby database to the primary database with the least interruption. Once the failover completes, AuDG creates a new standby database for you automatically, which will be a standby copy of your new primary Autonomous Database. If you wish to do a switchover, you can do so by clicking on the provided link on the **Autonomous Database Details** page, which we will see later as we progress.

In the case of a local standby database (in the current region), Autonomous Database does not give you access to the standby database.

Since the standby database is a copy of the primary database, all database features active in the primary instance will also be available whenever there is a failover or switchover from the primary to the standby instance. All operations performed by you in the primary – for example, scaling up the OCPU count or enabling auto scaling – are executed on the standby by AuDG. You can perform start and stop operations of the primary database, but you cannot perform any such operation on standby, as you do not get access.

In case of a local standby switchover or failover operation, where the standby takes the role of the primary, all database features active in the primary – for example, the OCPU count, auto scaling, licenses, display names, storage, database names, display names, tags, and license options – remain the same in the standby database post switchover or failover operation. OML Notebooks users and notebooks, apex configuration data and metadata, the access control list as set in the primary, and the private endpoint are exactly copied and made available in the standby database. The same Autonomous Database connection wallet is used after a failover or switchover operation. Thus, the end user applications, client connections, and APIs continue to work as before after a failover or switchover operation without requiring you to make any change to your existing application configuration.

The switchover is a manually triggered step; this can be initiated by clicking on the **Switchover** link in the **Autonomous Data Guard** section. The automatic failover operation is triggered based on the RPO and RTO.

Remote standby (cross-region standby)

When AuDG is enabled with cross-region, then it provisions a standby database of your primary Autonomous Database cross-region to protect your primary database from any disaster that entirely affects the cloud region where your primary Autonomous Database service is running.

However, there are certain limitations in terms of the regions where a replica standby database can be created and that entirely depends on the region where your primary Autonomous Database is created. The cross-regions where the standby database for your Autonomous Database can be created are called **AuDG paired regions**. These paired regions match the replication target regions of the OCI block volume primary region exactly. Hence, you must ensure that you have subscribed to the corresponding paired target region for your primary Autonomous Database region or the target region will not be listed when choosing a cross-region standby for your primary Autonomous Database. You can find the corresponding paired region by following this link:

```
https://docs.oracle.com/en-us/iaas/Content/Block/Concepts/
volumereplication.htm
```

The paired region is the value in the destination region mapped to the source region where your Autonomous Database is provisioned:

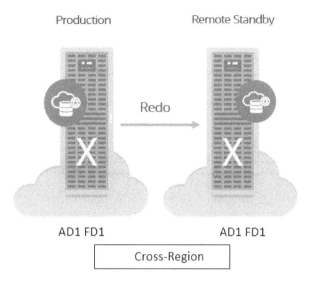

Figure 4.2: AuDG – cross-region standby

In contrast to local standby, where you enable remote standby using AuDG for your primary database, you can access the remote standby Autonomous Database from the OCI console. AuDG will perform all the operations that you perform in the primary Autonomous Database in the cross-region standby Autonomous Database – for example, enabling auto scaling or scaling up OCPUs on the primary. These operations are performed automatically in the cross-region standby Autonomous Database and hence you must ensure you have enough resources in the paired region, as in the primary region, for these operations to succeed.

Certain operations can be performed independently in the cross-region standby via the OCI console – for example, network configuration, creating VCNs for private endpoints, and tagging to define keys, values, and similar operations, which are not replicated between the primary and cross-region standby Autonomous Database.

Failover is handled differently in a cross-region standby. If the primary and local standby Autonomous Databases become unavailable, then you have to manually failover to the remote standby to take the role of the primary; AuDG does not perform an automatic failover to remote standby.

Although you can access the cross-region standby database in the OCI console, you can neither connect to the cross-region standby Autonomous Database nor can it be opened for read-only operations. It can only be connected when it takes up the primary role either due to a manual switchover or manual failover operation.

When you enable AuDG for a cross-region standby of your Autonomous Database, you have to download the wallet again after the standby creation. The wallet file contains connection strings for both the primary and standby Autonomous Database and the same wallet can be used for connecting after the failover or switchover to the remote standby Autonomous Database.

Since the order of connection strings in the wallet file impacts the connection time, you must use the wallet downloaded from the region wherever the database is running in the primary role.

There are differences in the way some of the features or options operate in a cross-region standby when compared to a local standby, which we will discuss here:

- **Display name suffixed with _Remote**: When you enable AuDG cross-region, then the standby Autonomous Database bears the name of the primary database suffixed with **_Remote**, as you can see in *Figure 4.3*:

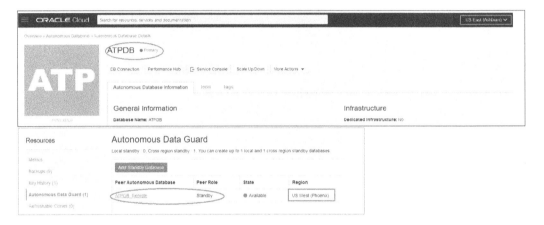

Figure 4.3 – Cross-region standby display name

Under **Resources**, when you click on the **Autonomous Data Guard** link on your **Autonomous Database Details** page, you can see the details of the standby database. The following are the definitions of the columns that we can see on this page:

- **Peer Autonomous Database**: This is the name of the Autonomous Database that is a replica.

- **Peer Role**: This shows the current database role of the peer Autonomous Database.

- **State**: This is the state of the peer Autonomous Database instance and it can be **Available** or **Stopped**.

- **Region**: This is **Peer Autonomous Database**'s region and will have the same region value as the replica instance if AuDG was enabled for a local standby – otherwise, it will have a different region value when AuDG is enabled for a cross-region standby.

In *Figure 4.3*, the name of the primary database is **Autonomous Transaction Processing Database (ATPDB)** with its role indicated as primary beside the name, and in the **Autonomous Data Guard** option, its Peer Autonomous Database name is the same as that of the primary database suffixed with **_Remote** to indicate that it's a cross-region standby database running in a different region from the primary region. **Peer Role** shows that it is a standby with an **Available** state for switchover or failover.

- **OML notebooks need re-creation**: The OML notebooks created in the primary region are not available in the cross-region post-switchover or failover. Hence, you must recreate the OML notebooks that were present in the primary region.

- **Private endpoint configuration in cross-standby**: The network configuration is not synchronized from primary to standby in a cross-region AuDG setup after a switchover or failover operation. Hence, you can independently configure and update private endpoints on the remote standby before failover or switchover. Additionally, you must complete the **Virtual Cloud Network (VCN) peering** and **domain forwarding** for wallets to work across regions. Autonomous Databases have private endpoints with AuDG enabled where primary and remote databases are in different VCNs.

- **APIs and scripts**: All APIs and scripts that were configured to manage your primary Autonomous Database must be updated to call APIs on a remote standby database region after a switchover or failover operation and it is recommended to use the wallet of the Autonomous Database wherever it runs with the primary role.

- **Client application connections**: No change to the application connection string is required either during a failover or switchover operation. The only recommendation is to use the wallet from the region where Autonomous Database is running with the primary role. This is because the order of the connection string in the instance wallet file will impact the database connection time. Since the same wallet file contains the entry for both the primary and standby databases, downloading it locally will mean that the primary database entry precedes the remote standby database entry.

- **Change in URLs for Autonomous Database tools**: The URLs of all the tools associated with Autonomous Database change after a switchover or failover operation to the cross-standby region. This is not impacted in the case of a local standby – that is, URLs of the different tools remain the same even after switchover or failover in the case of a local standby AuDG configuration. These tools URL changes include Database Actions, **APEX**, **ORDS**, the Service Console, OML Notebooks, user management, and Graph Studio.

In subsequent sections, you will learn about some of the associated terminologies that apply to an AuDG configuration, what the different operations that you can perform are, and what the different statuses indicate in the respective environment so that you know what the values indicate and what actions you can perform. Let's understand them individually.

Associated specifications in AuDG

Before we can go ahead with enabling AuDG, let's first understand the following terminologies that are associated with AuDG.

Region and roles specification

In a local AuDG configuration, the values of regions and roles for a primary and standby Autonomous Database do not differ, but the values carry a difference with a cross-region AuDG configuration. AuDG will assign the value for the region as either primary or remote based on its role at the time when you enabled the **Autonomous Data Guard** option. That is, if the database has the primary role, then the region value would be primary, and if the role is standby, then the region would be remote and this value for the database does not change after a switchover or failover:

- **Primary**: This is the region where the primary database or local standby database is provisioned

- **Remote**: This is the cross-region where the remote standby is initially created

During a failover or a switchover operation, only the value of the role for the Autonomous Database changes – that is, either primary or standby, but the region value does not change.

For example, if you perform a switchover operation from the primary database to the cross-region database, then the naming convention changes as follows:

- **Before switchover**: ATPDB has a **Primary** role and its region is **Primary** while **ATPDB_Remote** is its **Standby** database with a **_Remote** suffix and the region is **Remote**:

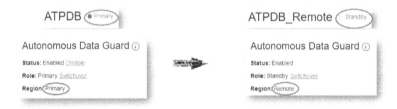

Figure 4.4 – Cross-region standby configuration – pre-switchover operation

- **After switchover**: The **ATPDB** role has changed to **Standby** as you can see in its name and its region is unchanged. This means that **ATPDB_Remote** now has the **Primary** role and its region is also unchanged – that is, it's the **Remote** region:

Figure 4.5 – Cross-region standby configuration – post-switchover operation

RPO and RTO are two Data Guard-specific concepts on which you build your DR strategies. Let's take a look at how both of these are defined in the context of an Autonomous Database.

RTO and RPO in an AuDG configuration

RTO is defined as the maximum amount of time required to restore the connection to the standby database after a manual or automatic failover. This is expressed in minutes and the value is usually representative of how much service downtime can be tolerated by the organization in case the database is not available.

RPO is the maximum duration of potential data loss of the primary that can be tolerated during recovery by an organization without causing any critical business loss.

In the context of AuDG, the RPO and RTO values have been pre-defined for local standby based on the type of failover that takes place:

- **Automatic failover**: This is an event when user sessions cannot connect to the primary instance service for a few minutes, which results in automatic failover to the standby database that takes over the role of primary, and the auto-failover of user sessions succeeds only when no data loss is guaranteed. In this case, the RTO is set to 2 minutes, and the RPO is set to 0 minutes.

- **Manual failover**: If for some reason, the automatic failover was unsuccessful, then you may initiate manual failover directly from the link on the **Autonomous Database Details** page. During a manual failover, the system automatically recovers the maximum amount of data, thus minimizing any potential data loss and it may vary from a few seconds to a few minutes of data loss. In the case of local standby, the RTO is 2 minutes and the RPO is 5 minutes.

This objective values differ with cross-standby. Automatic failover is not feasible in the current capability and with manual failover, the RTO is 1 hour and the RPO is up to 5 minutes.

Whatever the failover scenario may be, a new standby database is automatically provisioned for the new primary database post the failover operation.

Status and operation in AuDG

Whether an Autonomous Database is activated with AuDG or not is determined by the **Status** field in **Autonomous Data Guard** section of the **Autonomous Database Information** page. It would bear either of these two statuses:

- **Enabled**: This indicates that AuDG is enabled for the Autonomous Database
- **Disabled**: This indicates that AuDG is not enabled for the Autonomous Database

Figure 4.6 shows the details page of an ATPDB:

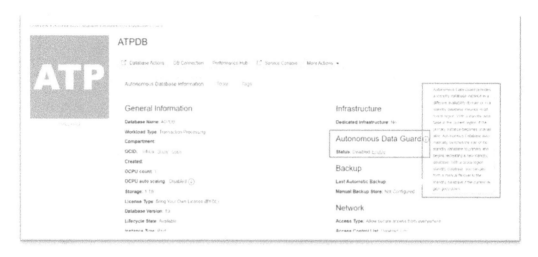

Figure 4.6 – ATP details page

If you notice in *Figure 4.4*, there is a section named **Autonomous Data Guard**, it has a status of **Disabled** and an **Enable** link beside that. If you click on the information symbol, it states the following:

"**Autonomous Data Guard provides a standby database instance in a different availability domain or in a standby database instance in different region. With a standby database in the current region, if the primary instance becomes unavailable, Autonomous Database automatically switches the role of the standby database to primary and begins recreating a new standby database. With a cross-region standby database, you can perform a manual failover to the standby database if the current region goes down.**"

That is a quick tooltip to inform anyone about the feature and how it performs failover to ensure high availability.

Now, let's discuss some of the operations associated with AuDG.

Disable operation

This allows you to disable AuDG for your Autonomous Database. There could be various reasons why you would like to disable AuDG. For example, you may wish to disable AuDG on a database instance that no longer serves as a production database or that doesn't have critical data anymore after a consolidation operation and can only be used for development. In this case, data loss can be tolerated and a regular backup would preserve the changes. There may be a scenario in which to have a cross-region standby and disable the local standby, instead of having both. It may be the organization strategy as per the business needs to disable AuDG in one region and enable it in another region and likewise, other reasons where you may feel the necessity to disable AuDG.

If AuDG is already enabled, then the **Disable** link appears against the **Status** field under the **Autonomous Data Guard** section of the **Autonomous Database Details** page. Note that disabling a remote standby is only allowed from the primary region. If you have performed a switchover or manual failover to the remote region, the database in the primary region will have a standby role. This will not allow you to disable the remote standby in the primary region. In order to disable the remote standby, first, you must perform a switchover so that the database in the primary region runs with a primary role. The **Disable** link will appear only if the Autonomous Database in the primary region has a primary role.

Click on the **Disable** link where you are prompted to select the autonomous standby database and provide the database name and confirm the **Disable** dialog box. If you have both a local and remote standby, you will have to disable each one of them individually. Disabling AuDG changes **Lifecycle State** to **Updating** and in the background, the standby database is being terminated. It also generates a **Disable Autonomous Data Guard** work request, which would be visible in **Resources** under **Work Requests**. If you enable AuDG for your Autonomous Database again, then it will create a fresh standby database for your Autonomous Database.

Enable operation

This allows you to enable AuDG for your Autonomous Database when **Lifecycle State** shows as **Available**. Click on the **Enable** link to configure AuDG for your Autonomous Database service. The Autonomous Database service could be an autonomous data warehouse, ATP, or an autonomous **JSON** instance. The **Enable** option will create a standby instance of your primary Autonomous Database. You can choose to either create a local or a cross-region standby database or both. Prior to enabling AuDG, you must ensure that you have ample resources (in terms of compute and storage) to provision a standby database, whether a local one, a remote one, or both. For a cross-region standby, your tenancy must be already subscribed to the paired regions. Otherwise, when enabling the cross-region standby, the remote regions will not be listed. During the process of enabling a cross-region standby, select the compartment where you have got the necessary IAM privileges. The **Enable** process creates an **Enable Autonomous Data Guard** work request that can be seen in the **Work Requests** menu option under **Resources**.

When you enable AuDG locally or cross-region, the Autonomous Database's **Lifecycle State** changes to **Updating**, and under the **Resources** area, the AuDG option shows the state of the standby databases as **Provisioning**. When **Lifecycle State** is **Updating**, then you cannot perform a move operation to another compartment or any stop, restart, and restore operation on the primary Autonomous Database. Later, after some time, the Autonomous Database's **Lifecycle State** will change to **Available** while the provisioning of the standby database goes on in the background. Once the provisioning of the standby database completes, then under the **Autonomous Data Guard** section of your **Autonomous Database Overview** page, the **Role** field will have a **Primary** value and a **Switchover** link for the standby, which you can click on to switch to the standby. If you enable a cross-region standby, then you will also see a **Region** field with a **Primary** value.

During the standby database creation process, your primary Autonomous Database is available for read or write operations and there is entirely no downtime incurred on your primary database, which is a key benefit, as business operations continue to function.

Post the creation of the standby database, you can check some of the attributes of the standby database by clicking **Autonomous Data Guard** under the **Resources** section. You can see the name of the standby database, which appears as a link under **Peer Autonomous Database**, its role under **Peer Role**, and the state and region of the standby database. In the case that your standby database is a cross-region standby database, you can access the peer Autonomous Database by clicking on the database link under the **Peer Autonomous Database** column, which opens the **Autonomous Database Details** page for the remote region.

Figure 4.7 shows the details page of the primary Autonomous Database instance showing a **Region** value of **Primary** and **Role** also as **Primary**:

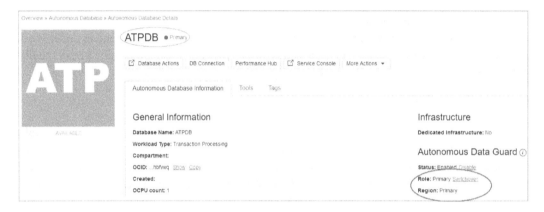

Figure 4.7 – ADB primary instance details

This has been configured with a cross-region standby as seen in *Figure 4.8*:

Figure 4.8 – Add Standby Database

When you want to add an additional standby database instance, you can do that from the primary database region and give the Autonomous Database a primary role. Note that you can add one local standby and one remote standby database only – no more than that.

As shown in *Figure 4.6*, you can add a new standby database by clicking on the **Add Standby** tab with the **Autonomous Data Guard** link under **Resources** on your primary **Autonomous Database Details** page.

Switchover operation

Once you have enabled AuDG, you will see the **Switchover** link in the **Role** field under the **Autonomous Data Guard** section of your **Autonomous Database Details** page. Clicking on that link performs a role transition – that is, **Primary** becomes **Standby** and **Standby** becomes **Primary** without any loss of data.

With a local standby AuDG configuration, the **Switchover** link appears in the **Role** field of the primary database when both the primary database and local standby database are available.

In cases where you only have a cross-region AuDG configuration, the **Switchover** link appears in the **Role** field of both the primary and remote database when both the primary and the remote standby databases are in an **Available** state. However, the switchover can be only initiated by clicking the **Switchover** link in the **Role** field of your cross-region standby. If you try to initiate a switchover by clicking on the **Switchover** link in the **Role** field of the primary database, then it will error out with a message as you can see in *Figure 4.9*:

Figure 4.9 – Cross-region switchover not allowed from the primary

When **Lifecycle State** for the primary Autonomous Database (primary region) is either **Available** or **Stopped**, and either the local standby or remote standby database is **Available**, then you can perform a switchover. The switchover operation can be initiated from the **Autonomous Database Details** page in the OCI console or you can invoke the Autonomous Database APIs. During a switchover operation, if there is any concurrent activity in progress on the Autonomous Database – for example, scaling up or manual backup creation – then you will be prompted to either pause or cancel the operation before it can proceed with the switchover operation.

Things to note that occur during a switchover operation have been outlined here:

- Switchover has to be started from cross-region standby only.

- Post-switchover, the state of the primary Autonomous Database is preserved – that is, if the primary was in a **Stopped** or **Available** state, then after the switchover operation is complete, the new primary database will also be in a **Stopped** or **Available** state.

- During a switchover operation, most actions on the database in the OCI console are disabled and the **Lifecycle State** value is **Updating**.

- The switchover process creates an Autonomous Database work request that can be checked under the **Resources** section within the **Work Requests** option.

- Post-switchover or a failover operation, all graphs displayed in the service console and the OCI metrics represent information about the new primary.

- When you have enabled AuDG with both a local standby and a cross-region standby and you have performed a switchover to the remote standby, this means that remote standby has taken the primary role. Here, AuDG does not provide a local standby that would be available. It is only after the Autonomous Database in the primary region gains back the primary role that the local standby is made available. Usually, the switchover to the remote standby is performed when you are performing some kind of testing.

- You cannot cancel a cross-region switchover operation once initiated and the **Lifecycle State** value shows as **Role Change in Progress**.

Since we have two different options for enabling standby with AuDG, let's understand the behavior of the switchover in two different combinations:

1. **Enabled with no local standby, but with a cross-region standby**: When you have enabled AuDG with only a cross-region standby, then you can switch over cross-region so that the primary region Autonomous Database takes up the standby role and the remote region standby database takes up the primary role, but note that the switchover operation must be initiated from the remote region standby, but not from the primary region.

2. **Enabled with a local standby and cross-region standby**: When you have enabled AuDG for both a local and cross-region standby, then you can switch over from a primary to a local standby in the same primary region so that the primary database becomes the local region standby and the standby becomes the local region primary database. AuDG continues to use the same remote standby cross-region.

But if you switch over from the primary database in the primary region to the cross-region standby database, you cannot switch over to the local standby database in the primary region.

A switchover from the remote region database is possible when the remote region standby database has the primary role and the primary region's primary database has the standby role.

Automatic failover operation

Failover of database instances can happen from the primary database to standby when the primary database becomes unavailable for some reason. In an AuDG with a local standby, the system monitors the condition of the primary and whenever the primary instance becomes unavailable, failovers from the primary to the local standby instance happen automatically based on the RTO and RPO defined, provided the local standby database is available. This means that standby takes over the role of primary. During an automatic failover, if the local standby database is available, then automatic failover would guarantee zero data loss as per the RPO target. After an automatic failover operation, a new local standby is automatically provisioned in the local region after the standby transitions to the primary role.

While the system is provisioning a new standby database after a failover operation to either a local or remote standby, AuDG is not enabled on the primary unless the provision is completed. After the provisioning is completed and the standby is available, AuDG is enabled on both the primary and new standby. Automatic failover is disabled when the **Lifecycle State** value is either **Restore in Progress** or **Upgrading**.

You can view the last failover operation details by hovering on the (i) icon in the **Role** field.

In certain scenarios where the primary becomes unreachable or has failed, the AuDG automatic failover conditions are not met, or the RPO target is not met, the OCI console displays a banner stating that the automatic failover did not succeed while mentioning the reasons and provides a link to initiate manual failover.

Manual failover operation

If automatic failover to your local standby did not succeed and the standby database is in an **Available** state, then you can perform a manual failover to make your local standby a primary database. If a cross-region standby is available, you can perform a manual failover to the remote standby. However, you must note that there is a possibility of data loss during a manual failover.

As in the case of a switchover, let's examine different scenarios and understand the behavior of a failover operation:

- **Enabled with no local standby, but with a cross-region standby**: In this configuration, since an automatic failover to a cross-region standby is not feasible, you have to initiate a manual failover to a cross-region standby when you detect that the primary database in the primary region is down. We have already discussed that the RTO is 1 hour and the RPO is up to 5 minutes in the case of a cross-region manual failover. When you initiate a switchover from a cross-region standby database and the operation fails, then AuDG will display a **Failover** link in the **Role** field of the cross-region standby that you can click to initiate a manual failover. You can also use APIs to initiate a manual failover.

- **Enabled with a local standby only**: When AuDG is enabled with a local standby in the primary region and a remote standby in the remote region and the primary database becomes unavailable, then AuDG will automatically failover to the local standby in the primary region with minimal interruption if the standby state is available. If the automatic failover is not successful, then the OCI console displays a banner with the information about why the automatic switchover operation did not succeed and provides a failover link in the **Role** field of the primary database that you can click to initiate the manual failover to the local standby. The failover link will appear only if the **Lifecycle State** value of the primary database is **Unavailable** and the **Lifecycle State** value of the local standby is **Available**. You can initiate a manual failover anytime using the APIs. The RTO and RPO value for a local standby would apply as discussed earlier. At the end of the failover to the local standby, AuDG creates a new local standby in the same primary region.

- **Enabled with a local standby and cross-region standby only**: When the primary database in the primary region is unavailable and both the local and remote standby databases are available, if the automatic failover to the local standby is not successful, then it is recommended that you first attempt a manual failover to the local region standby. There is no change to the cross-region standby and AuDG continues to use the same cross-region standby.

  If the local standby is unavailable or a manual failover to the local standby fails, then you can perform a manual switchover to the cross-region remote standby (initiated from the **Switchover** link in the **Role** field of the cross-region remote standby) and that will make the remote standby the new primary database cross-region. If that switchover operation fails, then AuDG will create a **Failover** link in the **Role** field of the remote standby, which you click to initiate a manual failover to the standby database. The RTO and RPO value for the manual failover to the cross-region remote standby will apply as discussed earlier.

  Unlike a failover operation to a local standby, AuDG does not create a new local standby when you manually failover the primary database to the cross-region remote standby database. Thus, the remote region database becomes a primary database but does not have a local standby.

  After a manual failover operation, you can view the information about any data loss in a message in the OCI console banner or by hovering over (i) in the **Role** field.

Termination operation

When you terminate an Autonomous Database that is enabled with AuDG then the standby database is automatically terminated. If you have both a local and cross-region standby, then both are terminated.

Add new standby

After you have enabled AuDG and have created your first standby Autonomous Database, you can add a new standby database by clicking on the **Add Standby Database** tab in the **Autonomous Data Guard** option under **Resources** as you can see in *Figure 4.10*:

Figure 4.10 – Add Standby Database

Adding the standby is a pretty simple step so next, let's understand the different states as the creation of the standby is in progress.

Peer Autonomous Database State

The state of the peer Autonomous Database is indicated by the **State** field in the **Autonomous Data Guard** option under **Resources** on the **Autonomous Database Details** page as shown in *Figure 4.11*. It may have any of the following values based on the current state:

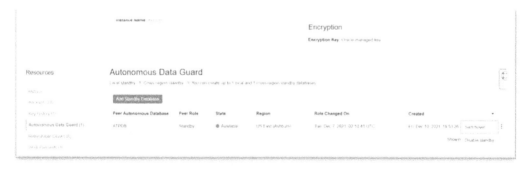

Figure 4.11 – Peer Autonomous Database state

- The **Provisioning** state is seen in the following scenarios:

  - The first time you perform an enable operation of AuDG, it will bear this status as it begins creating a standby database for your Autonomous Database. The state will change to **Available** once provisioning completes, which includes synchronization configuration as well.

- After failover to a local standby, you will see this state as well, since it recreates a local standby database.

- If the primary database is being restored from backup, in that case, a local standby database will be recreated – in which case, the state value will display as **Provisioning**.

- **Available** indicates that the standby database is available for a switchover or a failover operation

- **Role Change in Progress** indicates a switchover or failover operation is in progress

Thus, in this section of the chapter, we have come to understand some of the key concepts around AuDG that are essential to developing a good understanding of how DR works in an Autonomous Database service. In the next section, we will take a practical look via some screenshots that will equip us with the right information as to how we would go about configuring and using it.

Implementing AuDG

In this section of the chapter, we will browse through some of the screenshots that will give us a sufficient idea of how to configure AuDG for your Autonomous Database and perform certain operations that we have learned about so far.

Now, let's jump into the OCI console and start doing some of the operations. To begin with, let's enable AuDG for your ADB instance.

Enabling AuDG

As we understood earlier, you can enable either a local or a cross-region standby Autonomous Database with an AuDG configuration, but before you can enable AuDG for your Autonomous Database instance, you must ensure that you have enough OCPU and storage resources in your tenancy across the AD and region in which you want to host the standby ADB instance.

To enable AuDG, please follow these steps:

1. In order to enable AuDG, the primary instance's **Lifecycle State** should be **Available**, as indicated by **1** in *Figure 4.12*, on the **Autonomous Database Details** page. If you check on the section under **Autonomous Data Guard**, you can see the tooltip for a quick hint. The status shows as **Disabled**, which means AuDG is not enabled on this Autonomous Database instance as indicated by **2**:

Figure 4.12 – Pre-checks for enabling AuDG

2. Click on the **Enable** link to enable AuDG. This opens up a popup asking for confirmation as you can see in *Figure 4.13*:

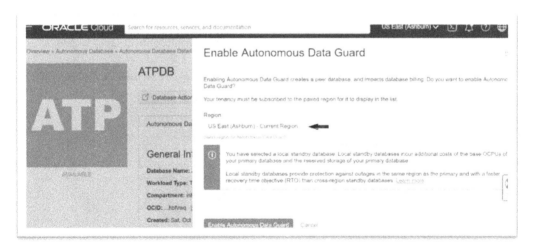

Figure 4.13 – Enabling AuDG locally

You can create a local standby by keeping the default region selected, which will be the same region as your Autonomous Database, and clicking on **Enable Autonomous Data Guard**.

3. If you wish to create a remote standby, then you can select another paired region from the dropdown that you have subscribed to as seen in *Figure 4.14*:

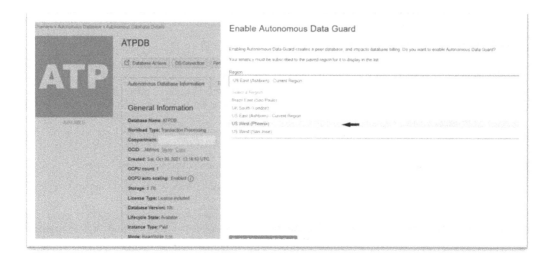

Figure 4.14 – Enabling AuDG cross-region

4.   When you select the region, you will be asked to select a compartment where you wish to host your remote standby as shown in *Figure 4.15*. You must ensure that in the subscribed region, you have a sufficient resource quota in terms of OCPU and storage resources and the needed permission to provision an Autonomous Database in the compartment cross-region:

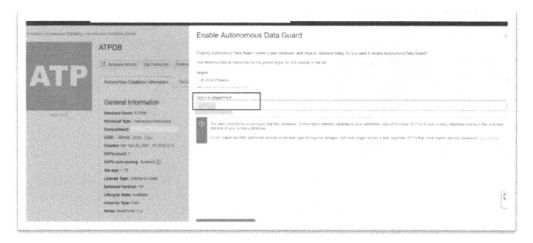

Figure 4.15 – Enabling AuDG cross-region – selecting a compartment

---

Important note

We will skip the steps to enable a cross-region standby and continue with enabling AuDG locally. The steps to create a cross-region standby can be initiated using the **Add Standby** process.

5. Now, click on **Enable Autonomous Data Guard** to begin creating a local standby as shown in *Figure 4.13*. This will change the state of your primary Autonomous Database to **UPDATING** as you can see in *Figure 4.16*:

Figure 4.16 – AuDG provisioning

6. Provisioning a local standby is pretty fast. The **Lifecycle State** value of the primary has changed to **AVAILABLE** as you can see in *Figure 4.17*. The **Autonomous Data Guard** section now shows the status of AuDG as **Enabled** with a link to disable AuDG beside it. The **Role** field shows the value of the role for this database in this region, which is **Primary**, and it is ready for a switchover as denoted by the active link for **Switchover**:

Figure 4.17 – AuDG provisioned

7. Now click on **Autonomous Data Guard** under the **Resources** section to check the details about the standby database as shown in *Figure 4.18*:

Figure 4.18 – Local standby peer

The **Peer Autonomous Database**'s name is the same as that of the primary database and **Peer Role** shows as **Standby** with an **Available** state in the same region as the primary Autonomous Database. Since the standby is **Available**, much like the primary, you can perform a switchover by clicking on the **Switchover** link in the primary as shown in *Figure 4.18*, or by clicking on the three dots in the row and then selecting **Switchover**. As mentioned earlier, the name in the **Peer Autonomous Database** field is not a link since AuDG does not allow access to the local standby.

8. Enabling AuDG will generate a work request, which you can see by clicking on the **Work Requests** option under **Resources** in the left-hand pane as shown in *Figure 4.19*. The state will change from **In Progress** to **Succeeded**:

Figure 4.19 – AuDG provisioning work requests

When the **% Complete** column has a value of **100%**, then it means that provisioning is complete and you will notice the primary database's **Lifecycle State**, as well as the peer Autonomous Database state, will be **Available**. In this way, we can create a local standby database for our Autonomous Database.

Let's also create a remote standby cross-region for your primary Autonomous Database:

1. Click on the **Autonomous Data Guard** option under the **Resources** section and click on the **Add Standby Database** link as shown in *Figure 4.20*:

Figure 4.20 – Adding a remote standby ADB

2. Adding a standby will open up a new window where you can choose a region other than the primary region where the remote database has to be provisioned. In this example, we chose the **US West (Phoenix)** region to deploy the remote standby database as shown in *Figure 4.21*:

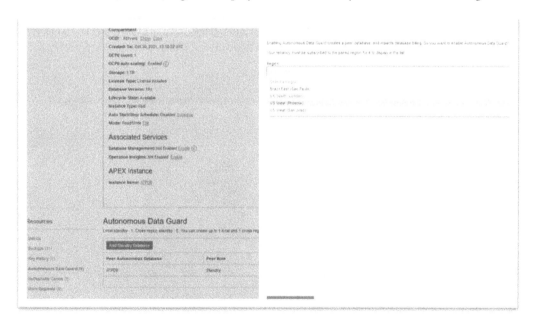

Figure 4.21 – Adding a remote standby ADB – selecting a region

This will lead to another selection where you need to choose the compartment for deployment in the paired region:

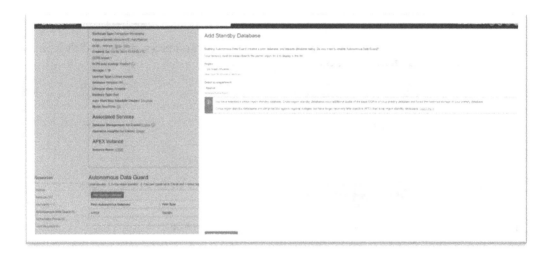

Figure 4.22 – Adding a remote standby ADB – selecting a compartment

3. Click **Add Standby Database**, which begins the provisioning of the remote standby database. During this step, the primary's **Lifecycle State** changed to **UPDATING**, and the state of the peer Autonomous Database changed to **Provisioning** as you can see in the highlighted box in *Figure 4.23*:

Figure 4.23 – Remote standby ADB – provisioning

This would also create a work request as it did for a local standby with the operation named **Enable Autonomous Data Guard**. You can check the **Lifecycle State** value for the primary and the peer Autonomous Databases. As seen in *Figure 4.24*, for the primary as well as for the remote standby, it's **Available**, which means the provisioning of the remote standby cross-region is complete:

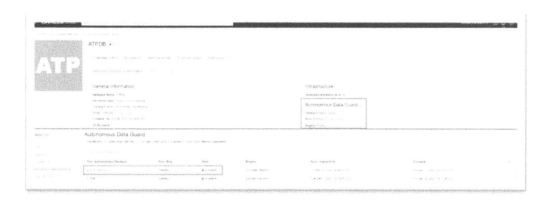

Figure 4.24 – Remote standby ADB – provisioned

4. If you note, the name of the peer Autonomous Database cross-region, that is, in the Phoenix region, is **ATPDB_Remote**, which appears as a hyperlink, but the local standby database, which bears the same name as the primary, is not a hyperlink. Clicking on **ATPDB_Remote** will open up a new tab providing details of the remote standby ATP database as shown in *Figure 4.25*, similar to the primary since AuDG allows access to the remote standby ATP database through the OCI console:

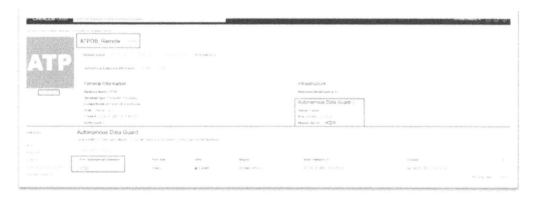

Figure 4.25 – Remote standby ATP in the OCI console

Note that in the remote region, since this involves provisioning a replica of the primary, there would be a work request created but with a different operation named **Create Autonomous Database** as you can see in *Figure 4.26*:

Figure 4.26 – ADB work requests for the remote standby

When you have enabled AuDG with a cross-region standby, here are a few additional points to keep in mind:

- If the primary database is restored from backup, a new remote region standby instance is created from the restored primary

- Automatic and manual backups can only happen from the database with the primary role

- The restore or clone operation from backup can only happen from the primary and you will have the option to restore when either the primary region or remote region database has the standby role

This completes the steps to be followed when enabling AuDG with a local and cross-region standby. The provisioning step is very simplified using the OCI console – you can provision a standby database for your Autonomous Database with just a few clicks. It is immediately ready for switchover and automatic or manual failover. It frees you of all the manual effort that you would have needed otherwise. In the next section, we will check a few switchover scenarios that you can self-initiate.

Performing a manual switchover operation – local standby

Typically, a database switchover operation is performed to evaluate the preparedness of the secondary database. It will resemble a failover scenario and you will be able to ascertain whether your application is appropriately configured to the automated action, as would be required during a database failover or switchover scenario or for audit or certification reasons. There may be other requirements within your organization to perform a switchover.

In an AuDG configuration, a switchover is initiated just by clicking on the **Switchover** link in the **Role** field. You will see the **Switchover** link when the primary database's **Lifecycle State** is **AVAILABLE** or **STOPPED** and the **Peer State** field is **Available**. Click on **Switchover** as indicated by the bold arrow in *Figure 4.27*:

Figure 4.27 – Initiating a switchover

Figure 4.28 shows you the option to choose local or remote:

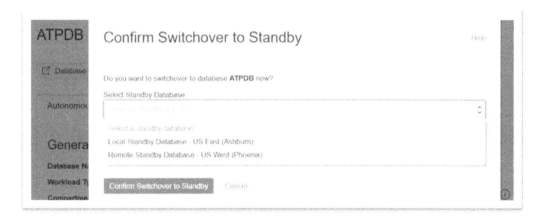

Figure 4.28 – Choosing a local or remote standby for switchover

If you choose remote, then you are notified that the operation is not supported as you can see in *Figure 4.29*. In the case of a remote standby, when a primary is available, you must switch over from a remote standby database:

Figure 4.29 – Choosing a remote standby from the primary – not supported

Since we have a local standby, we can now switch the primary over to a local standby while being in the primary by selecting the local standby in the dropdown and providing the name of the standby database as shown in *Figure 4.30*, and clicking **Confirm Switchover to Standby**:

Figure 4.30 – Switchover to local standby

Once initiated, the primary database **Lifecycle State** value changes to **UPDATING**, and the **State** field value of the peer Autonomous Database local standby changes to **Role Change in Progress**, as you can see in *Figure 4.31*:

Figure 4.31 – Role Change in Progress during local switchover

After the local standby takes the role of primary and the old primary takes over the role of standby, there is fresh provisioning happening for the remote standby cross-region. This can be seen in the **State** field of the peer Autonomous database as shown in *Figure 4.32*. If you check **Work Requests** in the remote Autonomous Database console, as indicated by **2**, the operation name is **Create Autonomous Database**, which means it is freshly provisioning the remote standby database to map with the new primary:

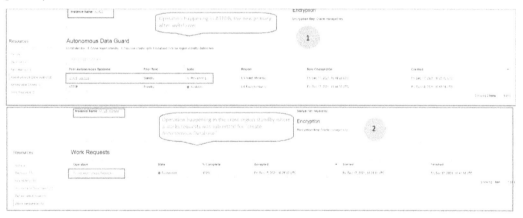

Figure 4.32 – Post-switchover to the local standby

All operations in the primary database, such as stop, restart, move, or restore, are restricted, along with other operations under **More Actions** on the **Autonomous Database Details** page, and you can only perform these operations after the switchover operation has been completed.

At the end of the local switchover operation, you can check the region of the Autonomous Database, which is **Primary** – that's because, although the role change has happened for the initial primary to stand by, the region of the database does not change. This is depicted in *Figure 4.33*:

Figure 4.33 – Local switchover completed

As we performed a local switchover, let's now see how to perform a switchover to a remote standby in the next section. The steps remain pretty much the same, except the underlying operations are a little different than what happens in case of a local switchover.

Performing a manual switchover operation – remote standby

Unlike how we performed a manual switchover from a primary region, the switchover operation in the case of a remote standby must be started cross-region. Click on the remote standby name in the **Autonomous Data Guard** option while in the primary region as shown in *Figure 4.34*:

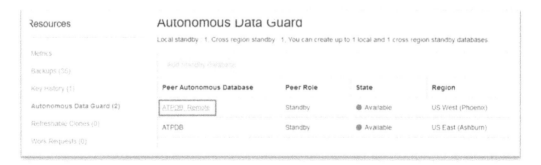

Figure 4.34 – Clicking on the remote ADB

This will open up a new tab for the remote Autonomous Database as you can see in *Figure 4.35*. The **Region** value is **Remote** since this standby has been created cross-region. The **Role** value is **Standby**. Click on the **Switchover** link to initiate the switchover to the primary:

Figure 4.35 – Initiating a remote switchover

This opens up a popup with the preselected region of the remote standby and asks you to confirm the name of the database before starting the switchover operation as shown in *Figure 4.36*. Fill in the name of the ADB and click on **Confirm Switchover to Standby**:

Figure 4.36 – Confirming a remote switchover

This would change the **Lifecycle State** value of the database in the remote region to **ROLE CHANGE IN PROGRESS** and that of the primary to **Updating**. *Figure 4.38* shows the status as observed cross-region:

Figure 4.37 – Lifecycle state during the switchover to remote

Figure 4.38 shows the given lifestyle state for the primary region before the switchover is complete:

Figure 4.38 – Lifecycle state of the primary region during switchover

After some time, **Lifecycle State** for **ATPDB_Remote** is **AVAILABLE**. Additional information that you must note is that the role has now changed to **Primary** but the **Region** value does not change. It is still the **Remote** region as per the initial configuration. You can view these changes in the console for the cross-region standby as in *Figure 4.39*:

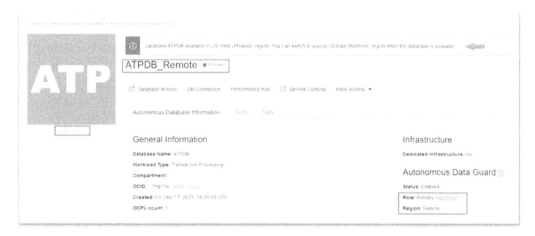

Figure 4.39 – Primary role in a remote region

ATPDB in the primary region will have a change in status from **updating** to **provisioning** and post-completion of provisioning operation in the primary region, observe the new change in the role of the old primary, **ATPDB**. It has a standby role but the **Region** value remains unchanged as per the original configuration, which you can check in the **Autonomous Data Guard** section in *Figure 4.40*:

Figure 4.40 – Standby role in the primary region

As you may have noticed, in *Figure 4.39* and *Figure 4.40*, the **DB Connection** tab is disabled in the primary region (**ATPDB** – running as standby) and is active in the remote region where **ATPDB_ Remote** is running with a primary role. This will allow you to download the wallet for connecting to the database as you would do in order to connect to Autonomous Database in general.

In the case of a cross-region switchover, when the remote region standby becomes the new primary, it won't have a local standby in the remote region, as you can see in *Figure 4.41* in the **Autonomous Data Guard** option of **ATPDB_Remote**, which now has a primary role. It states **Local standby : 0** and **Cross region standby : 1**:

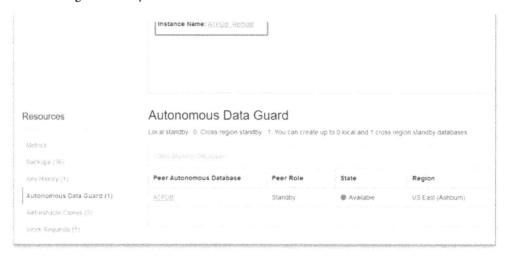

Figure 4.41 – No local standby in the remote primary

When you switch over the primary database from the remote region to the primary region, as part of the switchover operation, it will first perform a switchover from the remote region to the primary region. Then, it provisions a fresh local standby that was present in the initial configuration as you can see in *Figure 4.42*:

Instance Name: ATPDB

Resources

Metrics

Backups (20)

Key History (1)

Autonomous Data Guard (2)

Refreshable Clones (0)

Work Requests (2)

Autonomous Data Guard

Local standby 1. Cross region standby 1. You can create up to 1 local and 1 cross region standby databases

Peer Autonomous Database	Peer Role	State	Region
ATPDB_Remote	Standby	Unavailable	US West (Phoenix)
ATPDB	Standby	Provisioning	US East (Ashburn)

Figure 4.42 – Switchover from a remote to a primary region

As the next step, it also provisions the remote standby database in the same cross-region as was present in the initial configuration. The **Unavailable** state for **ATPDB_Remote** as we saw in the previous screenshot changes to **PROVISIONING**, which you can see in *Figure 4.43*, and finally, the state will change to **Available** once the provisioning of remote standby has been completed. This will set the configuration to the initial one that existed before the switchover:

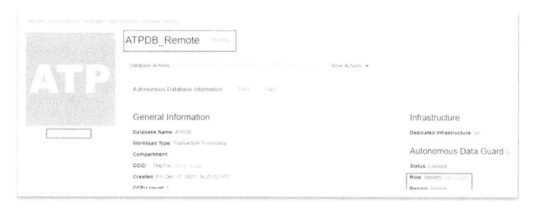

Figure 4.43 – Provisioning a remote standby post-switch back to the primary region

Each of the switchover operations we tried will create work requests as they did during the enabling of AuDG.

Thus, we have seen how to perform a switchover operation for your primary to either the local or cross-region standby with a few simple clicks, entirely overcoming all the manual steps that you need to perform this sequentially, which is more prone to errors unless you are an experienced DBA.

> **Important note**
>
> We have skipped the failover operation practice, as we don't have sufficient control over the Shared Autonomous Database infrastructure to imitate a manual failover condition, but the steps remain pretty straightforward as has been explained in the previous sections.

Although enabling AuDG is highly recommended for DR, there could be a scenario where one of the Autonomous Databases no longer holds any critical data and data loss can be tolerated in the event of a disaster. In this scenario, you can disable AuDG in the simplest way possible, which we are going to see in the next section.

Disabling AuDG

Disabling AuDG with a cross-region standby can only be initiated from the primary region. When the Autonomous Database is available and bears a primary role in the primary region, then the **Disable** link will appear active as you can see in *Figure 4.44*. Click on the **Disable** link to disable the standby database:

Figure 4.44 – Disabling AuDG

If you have a local and remote standby, then you have to repeat the **Disable** operation twice, selecting the specific standby database in each iteration, which eventually terminates both the standby databases of the Autonomous Database.

Clicking **Disable** opens up a popup prompting you to select the standby type and also asking you to enter the name of the Autonomous Database for confirmation as seen in *Figure 4.45*. Here, I have selected **Remote Standby Database - US West (Phoenix)** to disable:

Figure 4.45 – Choosing the standby to disable

Click **Disable Autonomous Data Guard**, which will change the status of the primary database to **Updating** and will terminate both the local and remote standby database as you can see in *Figure 4.46*:

Figure 4.46 – Disabling the remote standby

At the end of this remote **Disable** operation, it will freshly provision the local standby and make both the primary and local standby available for access as you can see in the following screenshot in *Figure 4.47*:

Figure 4.47 – Disabling the remote standby completed

As with the other operations, **Disable Autonomous Data Guard** follows a few simple steps to clean up the AuDG deployments for your Autonomous Databases that no longer need the configuration. This may be the case where you have moved the primary data to a new Autonomous Database and wish to keep this database for development with all sensitive data being completely masked.

During the disable operation, there is no downtime incurred and you can continue to use your primary Autonomous Database.

AuDG also provides APIs, SDKs, or a CLI that you can use to perform all the operations, such as enabling or disabling AuDG or performing a manual switchover or failover operation. Hence, you have the flexibility to use the tool of choice.

There are a few things to note here:

- As in other operations, during the disable operation, you won't be able to perform most of the options under the **More Actions** tab on the **Autonomous Database Details** page

- As with other operations, here too, the Autonomous Database generates a work request named **Disable Autonomous Data Guard**, which you can find under the **Resources** tab in the left-hand pane as shown in *Figure 4.48*:

Figure 4.48 – Disabling an AuDG work request

As with every other service in OCI, you can leverage **OCI Events Service** to generate notifications for AuDG events and alert respective teams on the occurrence of these events. Let's understand a few details about it in the following section.

Leveraging OCI Events Service for AuDG operations

You can configure OCI Events Service for your AuDG operations to be notified of specific operations such as enabling, disabling, manual failovers, switchovers, and specifying rules in response to these events. Take a look into the OCI Events Service console and create a rule for a specific event type, a snapshot of which is shown in *Figure 4.49*. Look for the list of event types related to the AuDG operation and set the rule with an appropriate action, which could be an email notification, for example.

I am not going into much detail on OCI Events Service, as it is not the core topic here, but to facilitate your AuDG operation, I wanted to highlight the pre-defined AuDG events available that can definitely simplify your management of an Autonomous Database configured with AuDG:

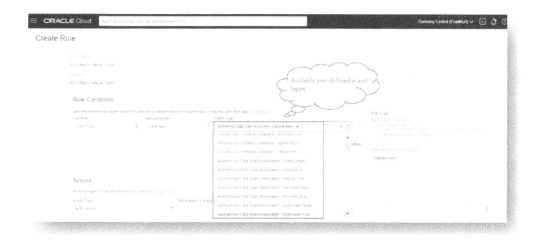

Figure 4.49 – OCI event types for AuDG

There are few features regarding autonomous data guard configuration that you must note, which I will highlight in the following section.

Points to remember about AuDG

Here are a few general notes to remember before we wind up on the topic of AuDG:

- While the disable operation is running, that is, **Lifecycle State** for the primary shows as **Updating** and the peer Autonomous Database **State** field shows as **Terminating**, there is no downtime required for the primary – you will still be able to perform your normal database operation on the primary Autonomous Database while the disable operation is in progress.

- You cannot connect to a local standby to offload read-only queries.

- You can access the **Autonomous Database Details** page for a remote standby database, but you cannot open it in read-only for running read-only queries.

- Wallets are not accessible in the remote standby database. Hence, you cannot rotate a remote standby database wallet. You must switch over from the remote standby in order to enable access to the wallet.

- Only Oracle-managed encryption keys are supported by cross-region standby databases, which means you cannot change from Oracle-managed encryption keys to customer-managed encryption keys if your Autonomous Database has a cross-region standby. If the Autonomous Database already uses customer-managed keys for encryption, you cannot enable a cross-region standby.

- The CPU utilization and OCPU's allocated graph or any metrics of this kind only apply to the primary and do not apply either to the local or remote standby databases.

- Automatic failover to a local standby is disabled when there is either an upgrade or a restoration operation ongoing.

- When the **Lifecycle State** value for the primary Autonomous Database is **Stopped**, then the **Lifecycle State** value of the standby is also **Stopped**, but you may still perform a switchover when the primary is **Stopped**.

- An operation such as scaling up or enabling auto scaling performed on the primary database is automatically executed by AuDG on the standby side for both a local and a remote standby. There is an exception here – in the case of a remote standby, a few additional operations, such as private endpoint configuration, can be done in the remote standby independent of the primary.

- You must reset and verify the access control list on the remote standby database after you disable the private endpoint in the primary.

- **Oracle Data Safe** cannot be enabled on an Autonomous Database that has the cross-region standby enabled.

- AuDG is not available with the Always Free subscription.

- All AuDG operations can be performed using the AuDG APIs.

- You can only allow **TLS connections** on the primary if the primary and remote standby are both configured to support the TLS connection.

- AuDG is also now configurable with an Autonomous Database on Dedicated **Exadata infrastructure** and supports intra-region and cross-region DR capability, but unlike Autonomous Database on Shared infrastructure, AuDG on Dedicated infrastructure is enabled at the container level and the option is checked while provisioning the container database. Hence, when creating the Autonomous Database, you must choose those container databases that were enabled with AuDG during provisioning time.

In the case of an Autonomous Database on Dedicated infrastructure that is enabled with AuDG, you have the option to reinstate the database after a failover. This concludes the section about configuring AuDG in an Autonomous Database environment and also brings us to the end of the chapter on AuDG.

Summary

In this chapter, we got some insights into how Autonomous Database provides another layer of protection and high availability by enabling either a local or cross-region standby using the fully automated feature called AuDG. We came to understand the different concepts related to the AuDG operation and the different status changes it goes through when we perform different operations.

With AuDG, it was also interesting to learn about the different switchover and failover scenarios and how the operation proceeds either automatically or with manual intervention required during a cross-region failover. It was also suggested to use the appropriate wallet while establishing a connection to the primary database in a respective region. Please consider some hands-on practice to grasp the underlying concepts better.

If you have an interest in carrying out operations using the APIs or SDKs, you can refer to the public documentation for the AuDG APIs so that you can automate some of the actions by invoking them remotely. This marks the end of this chapter, so now let's do a knowledge check of all that you have learned from this chapter.

Questions

1. When the primary database in the primary region becomes unresponsive, you can perform a failover from the cross-region standby from the primary directly from the console. State true or false.

2. With an AuDG configuration with both a local and remote standby, when you switch over from the primary to a remote standby, an automated backup will be initiated in the remote region database with the primary role. State true or false.

3. What are the RTO and RPO of a local standby during automatic failover?

4. If you have enabled AuDG with a cross-region standby and you enable auto scaling in the primary region's primary database, then you have to manually enable auto scaling in the other region's remote standby database. State true or false.

5. You can mask sensitive data in the cross-region remote standby. State true or false.

Answers

1. False. You have to perform a switchover first; if it fails, then it'll provide a **Failover** link in the console.

2. True. Automatic backups will all be initiated from the database with a primary role in an AuDG configuration.

3.   RTO = 2 mins, RPO = 0 mins.

4.   False. You only enable it in the primary and AuDG will automatically execute the command in the standby. Only certain configurations can be carried out in the standby.

5.   False. You cannot enable Oracle Data Safe when you have already configured a cross-region standby.

Further reading:

- https://docs.oracle.com/en/cloud/paas/autonomous-database/adbsa/part-using.html

- https://docs.oracle.com/en/cloud/paas/autonomous-database/dedicated/adbau/index.html#articletitle

5
Backup and Restore with Autonomous Database in OCI

Backup is a crucial activity when we deal with any database. It derives the mechanism of how to back up the data either with the same or a different optimized format. If data loss happens due to a disaster or human mistake, then the data can be recovered only through a backup. We have many backup methods for Oracle databases, such as Data Pump, **Recovery Manager** (**RMAN**), and so on. In this chapter, we are going to discuss backup strategies in Autonomous Database.

In general, Oracle database backup can be taken either as full database backup having user data along with its database metadata (for example, RMAN backup), or we can extract only the user data (for example, Data Pump) and create a dump file. You may wonder how the backup can be taken for an autonomous database. We'll cover that in detail in this chapter.

We will be covering the following topics:

- Understanding backup types
- Backup through the OCI portal
- Backup through the command line
- Restoring database backup

By the end of this chapter, we will have an understanding of all available backup methods for an autonomous database.

Technical requirements

This chapter requires an understanding of traditional backup methods, Data Pump, and RMAN. Before trying out the techniques and steps mentioned in this chapter, we need an OCI account with privileges to create an Object Storage Bucket and an autonomous database, whether an Autonomous Transaction Processing (ATP) database or an Autonomous Database (ADB).

Understanding the backup types

In an autonomous environment, all DB tasks are managed automatically; in the same way, the backup will also be taken care of by itself. But we do have a provision to back up the data manually. Let's discuss both methods.

Working with automated backup in the OCI portal

Automated backup is configured by default when we create an autonomous database. It is a mandatory backup taken daily for autonomous databases. No manual effort is required in this backup method.

Backups

Backups are automatically created daily.

Create Manual Backup

Display Name	State	Type	Encryption Key	Started	Ended	▼
Jun 27, 2022 01:22:47 UTC	● Active	Auto Backup	Oracle-managed key	Mon, Jun 27, 2022, 01:14:47 UTC	Mon, Jun 27, 2022,	Restore
Jun 26, 2022 00:35:00 UTC	● Active	Auto Backup	Oracle-managed key	Sun, Jun 26, 2022, 00:30:08 UTC	Sun, Jun 26, 2022, (	Create Clone ⋮
Jun 12, 2022 13:10:35 UTC	● Active	Auto Backup	Oracle-managed key	Sun, Jun 12, 2022, 13:05:14 UTC	Sun, Jun 12, 2022, 13:10:36 UTC	⋮
Jun 11, 2022 14:43:07 UTC	● Active	Auto Backup	Oracle-managed key	Sat, Jun 11, 2022, 14:33:54 UTC	Sat, Jun 11, 2022, 14:43:07 UTC	⋮
Jun 04, 2022 18:32:50 UTC	● Active	Auto Backup	Oracle-managed key	Sat, Jun 4, 2022, 18:27:29 UTC	Sat, Jun 4, 2022, 18:32:50 UTC	⋮
May 19, 2022 01:24:24 UTC	● Active	Auto Backup	Oracle-managed key	Thu, May 19, 2022, 01:14:26 UTC	Thu, May 19, 2022, 01:24:24 UTC	⋮

Figure 5.1 – Automatic backup list

In general, an Oracle database has various methods to take backups. RMAN is one of the proven methods to take Oracle database backup. This method has the luxury of taking a backup of either the full database or only changes that happened after the last full backup (incremental backup). Also, it has features to validate the backup before **Restore**. In an autonomous environment, the automated database backup takes RMAN full backups weekly and incremental backups daily into the object storage. The backup will be retained for 60 days. The RMAN backup piece will get deleted automatically once it crosses the retention period. Since the automatic backup is managed by the database itself, we can't stop or change the timing of the automatic backup. The backup algorithm decides the appropriate time for the backup based on the database workload. Another specialty of this backup method is that an automatic backup will be taken daily even if the database is in a closed status. We can restore the database backup to a point in time using this automated backup. We will discuss the Restore topic later in this chapter.

Learning how to take a manual backup

The previous section talked about automated backup, but at the same time, a manual backup for autonomous databases can be taken. Mostly, a manual backup will be required before or after the execution of sensitive tasks. The manual backup can be taken through the OCI portal or by executing the command-line `expdp` utility. We will discuss both methods in detail.

Manual backup through the OCI portal

Oracle Autonomous Database provides the option to make database backups manually. This backup is going to be an RMAN full backup and it will be taken directly into object storage. We have two prerequisites to initiating this manual backup. It is a one-time activity:

- Define Cloud credentials in the database using the DBMS_CLOUD package
- Create an object storage bucket

Let's go through the prerequisite tasks in detail.

Generating Cloud credentials

In Cloud databases, we should create Cloud credentials at the database level to access the Cloud resources. The credentials can be created using the DBMS_CLOUD.CREATE_CREDENTIAL package, available in the database:

```
BEGIN
DBMS_CLOUD.CREATE_CREDENTIAL(
credential_name => 'Credential_Name',
username => 'Cloud_UserName',
password => '<authorization_token>'
);
END;
/
```

Here, the variables are as follows:

- `Credential_Name`: User-friendly name for credentials
- `Username`: Cloud login username
- `Password`: Authorization token created using login credentials

You can follow these steps to generate an authentication token:

1. Log in to the Oracle Cloud portal. Click on **Identity | Users**. It will show the user details page as shown. In the bottom left, under **Resources**, click on the **Auth Tokens** option. We can see the **Auth Tokens** page here:

Figure 5.2 – User auth tokens

2. Click on the **Generate Token** button, which will bring up the following page:

Figure 5.3 – Generate Token

3. Provide a meaningful description and click on the **Generate Token** button again. It will generate a token as shown:

Figure 5.4 – User token

4. Copy the token information in a text file. Don't forget to copy the token since it won't be shown again. This token will be used for the `password` column in the `DBMS_CLOUD.CREATE_CREDENTIAL` procedure:

```
BEGIN
DBMS_CLOUD.CREATE_CREDENTIAL(
```

```
credential_name => 'Credential_Name',
username => 'Cloud_UserName',
password => '<authorization_token>'
);
END;
/
```

5. Execute the procedure to generate Cloud credentials at the database level. After creating the credentials, make it as default in the database.

    ALTER DATABASE PROPERTY SET DEFAULT_CREDENTIAL = '<Schema Name>.<Credential Name>';

    An example is shown here:

    ALTER DATABASE PROPERTY SET DEFAULT_CREDENTIAL = 'ADMIN. OBJSTORE_CRED';

At this stage, we have created a default Cloud credential at the database level. The next prerequisite is creating an object storage bucket.

Creating an object storage bucket

We will continue with the following steps to create an object storage bucket:

1. Log in to the Oracle Cloud portal. Under **Core Infrastructure**, click on **Storage** and then click on **Buckets** from **Object Storage & Archive Storage** as shown:

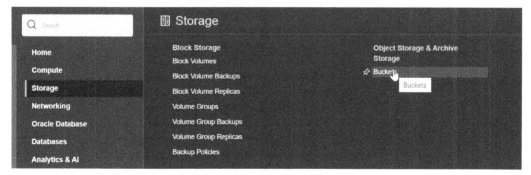

Figure 5.5 – OCI Object Storage

2. It will invoke the object storage bucket creation window as shown. Provide a bucket name and for **Storage Tier,** select **Standard**. **Archive** lets you create Archive buckets where archived objects will be stored and the storage cost is cheaper compared to a Standard bucket. Retrieving

objects from Archive buckets will take time, hence we choose the Standard bucket to store the database backup. We can also choose the other available options as per our requirements:

- **Enable Auto-Tiering** – Infrequently accessed backups will be moved from the Standard bucket to the Archive bucket.

- **Enable Object Versioning** – Objects will get a version number when the object is overwritten or deleted.

- **Emit Object Events** – Emits certain events based on object state changes. For example, if a new file is placed in object storage, it can send notifications or invoke functions through events.

- **Uncommitted Multipart Uploads Cleanup** – Deletes uncommitted or failed multipart uploads after 7 days.

Choose the appropriate **Encryption** method and then click on the **Create** button:

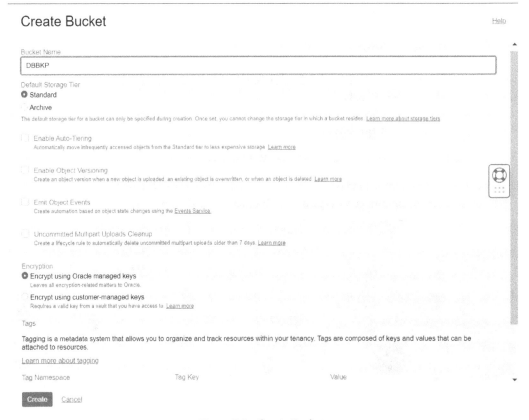

Figure 5.6 – Create Bucket

3.　We could make a default object storage bucket to store the autonomous database by setting the `DEFAULT_BACKUP_BUCKET` database property to the bucket name.

The bucket name should have the format of `https://swiftobjectstorage.region.oraclecloud.com/v1/object_storage_namespace/bucket_name`.

Here is an example:

```
ALTER DATABASE PROPERTY SET default_backup_
bucket='https://swiftobjectstorage.us-phoenix-1.
oraclecloud.com/v1/mynamespace/backup_database';
```

In this example, the object storage namespace is `mynamespace` and the bucket name is `backup_database`. The `namespace` value can be retrieved from the object storage bucket details page.

The database could be either an **ATP** database or an **ADB**. These prerequisites should have been created before invoking the manual backup.

Creating a manual backup

At this stage, all the prerequisites are set and ready, and we can execute the manual backup. The steps are as follows:

1. Log in to the Oracle Cloud portal and move either **ATP** or **ADB** to **Autonomous Database**. Click on the **Backups** option available in the **Resources** section:

Figure 5.7 – On-demand backup

2. Click on the **Create Manual Backup** button. It will present a pop-up window that specifies **Display Name** for the backup as shown:

Figure 5.8 – Create Manual Backup

3. After providing the display name, click on the **Create Manual Backup** button. The backup process will get initiated and the progress, along with the status, will be shown in the portal:

Backups

Backups are automatically created daily.

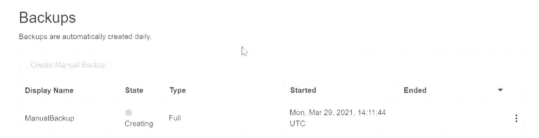

Figure 5.9 – Backup creation in progress

When the manual backup is in progress, some management activities, such as **Stop** or **Restart**, will get disabled. Also, the **ATP** banner changes from a green color to a yellow color, as shown in the following screenshot. Once the backup is complete, the status will change to **Available**, the color will change back to green, and all the functionalities will become active.

Figure 5.10 – ADB options are disabled

The backup status changes to **Active** once it is completed. The backup will be stored in object storage, but we will not be able to see or download the backup piece as the backup is managed automatically. The backup piece will be stored in the object storage for 60 days.

If there is a need to rename the autonomous database, there is no need to worry about existing backups. That will continue to work.

Once the backup is complete, we get two options, as shown in the following screenshot. Either the full backup can be restored or a clone database can be created.

Figure 5.11 – Restore backup

We will now move on to restoring backups.

Restoring database backups

There could be situations where unwanted DML or DDL changes happened in the database that cannot be rolled back, or the database may have been corrupted and needs to be repaired. In these case, the backup can be restored. In Autonomous Database, restoring is very simple. It is done through a single click.

The option to restore a database backup is available under the **More Actions** button on the **Autonomous Database Information** page, as shown:

Figure 5.12 – Database options

Choosing the **Restore** option will invoke the **Restore** window, as shown:

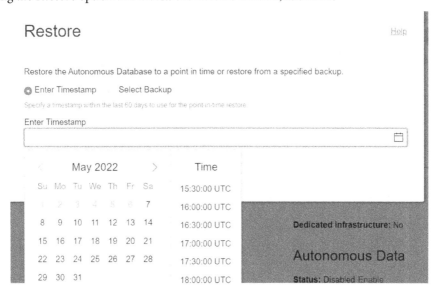

Figure 5.13 – Restore database backup

Point-in-time recovery can be done either at the preferred timestamp in UTC or any specific backup, as shown:

Figure 5.14 – Point-in-time restore

Alternatively, you can select a backup from the list:

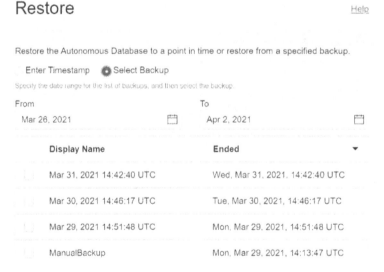

Figure 5.15 – Backup list

Choose the appropriate method and click **Restore**.

You can also restore from the backups listed on the **Autonomous Database** details page. You can find the list of available automatic and manual backups in the Cloud portal. You can choose the desired backup piece and click on the **Restore** option, as shown:

Figure 5.16 – Restore a particular backup

It will ask for confirmation before proceeding with the actual restore operation:

Figure 5.17 – Restore confirmation

By clicking on the **Restore** button, we confirm the restoration. The **ATP** status changes to **RESTORE IN PROGRESS** and the basic functionalities will be disabled until the restoration is completed:

Figure 5.18 – Restore in progress

We can monitor the Restore progress through **Work requests**:

Figure 5.19 – Work requests

Once Restore is completed, the banner changes from yellow to green, the status changes to **Available**, and all portal options will get enabled.

Manual database backup using Data Pump

Data Pump is a powerful tool to take logical backup and it has been introduced in Oracle Database 10g. It extracts the user data with its metadata and generates a dump file. That dump file can be imported into the target database. In general, it is a server-side utility, which means the dump will get stored in the database server. In an autonomous environment, we won't get server access, so how we will take a backup? The answer is either to create an autonomous dump set and then move it to object storage, or directly take the dump to object storage. We will discuss both methods here.

Creating a dump set

The step for creating a dump set is similar to the on-premises environment. First, we need to create a logical directory at the database level to store the dump:

```
SQL>  create directory dump_dir as 'datapump_dir';
Directory created.


SQL> select directory_name , directory_path from dba_
directories;


DIRECTORY_NAME


DIRECTORY_PATH
--------------------------


------------------------------------------------------------
--------------------------------
DUMP_DIR


/u03/dbfs/BDF28FA2296E84A6E053C614000ACEA2/data/datapump_dir
```

The /u03/dbfs/BDF28FA2296E84A6E053C614000ACEA2/data/ folder is the root location in the autonomous environment, and all custom directories will get created here. Remember, we can't access these folders physically.

The Data Pump expdp command is as follows:

```
expdp Admin/password@ATPDB_high
directory=dump_dir
dumpfile=exp%U.dmp
parallel=16
encryption_pwd_prompt=yes
filesize=1G
logfile=export.log
```

Here, encryption_pwd_prompt encrypts the dump using the given password. The same password should be used during import.

Here is an example:

```
$ expdp admin@kkatp_high filesize=1GB dumpfile=exp%U.
dmp  parallel=2 encryption_pwd_prompt=yes logfile=export.log
directory=dump_dir schemas=test
```

```
Export: Release 12.2.0.1.0 - Production on Fri Apr 2 12:19:20
2021

Copyright (c) 1982, 2017, Oracle and/or its affiliates.  All
rights reserved.
Password:

Connected to: Oracle Database 19c Enterprise Edition Release
19.0.0.0.0 - Production
Encryption Password:
Starting "ADMIN"."SYS_EXPORT_SCHEMA_01":  admin/********@kkatp_
high filesize=1GB dumpfile=exp%U.dmp parallel=2 encryption_pwd_
prompt=yes logfile=export.log directory=dump_dir schemas=test
Processing object type SCHEMA_EXPORT/TABLE/TABLE_DATA
Processing object type SCHEMA_EXPORT/TABLE/STATISTICS/TABLE_
STATISTICS
. . exported "TEST"."TEST"                          8.568
MB 1000000 rows
Processing object type SCHEMA_EXPORT/STATISTICS/MARKER
Processing object type SCHEMA_EXPORT/USER
```

```
Processing object type SCHEMA_EXPORT/DEFAULT_ROLE
Processing object type SCHEMA_EXPORT/TABLESPACE_QUOTA
Processing object type SCHEMA_EXPORT/PASSWORD_HISTORY
Processing object type SCHEMA_EXPORT/PRE_SCHEMA/PROCACT_SCHEMA
Processing object type SCHEMA_EXPORT/TABLE/TABLE
Processing object type SCHEMA_EXPORT/POST_SCHEMA/PROCACT_SCHEMA
Master table "ADMIN"."SYS_EXPORT_SCHEMA_01" successfully
loaded/unloaded
******************************************************************
***************
Dump file set for ADMIN.SYS_EXPORT_SCHEMA_01 is:
  /u03/dbfs/BDF28FA2296E84A6E053C614000ACEA2/data/datapump_dir/
exp01.dmp
  /u03/dbfs/BDF28FA2296E84A6E053C614000ACEA2/data/datapump_dir/
exp02.dmp
Job "ADMIN"."SYS_EXPORT_SCHEMA_01" successfully completed at
Fri Apr 2 12:20:00 2021 elapsed 0 00:00:26
```

The dump has been stored in /u03/dbfs/BDF28FA2296E84A6E053C614000ACEA2/data. We don't have access to the database server, but the dump file existence can be verified through a query:

```
SQL> SELECT * FROM DBMS_CLOUD.LIST_FILES('DATA_PUMP_DIR');
OBJECT_NAME

BYTES
CHECKSUM
CREATED




LAST_MODIFIED
---------- -------------------------------------------------------
--------------------------------------------------------------
exp01.dmp

45056
```

```
02-APR-21 11.36.59.000000 AM +00:00


02-APR-21 11.37.57.000000 AM +00:00
exp02.dmp

9297920


02-APR-21 11.37.25.000000 AM +00:00

02-APR-21 11.37.57.000000 AM +00:00
```

Now, the dump files need to be moved to object storage. First, we need to create an object storage bucket:

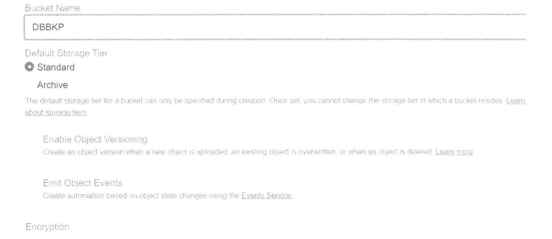

Figure 5.20 – Creating an object storage bucket

Now we need to copy the dump files to object storage. This can be achieved by executing the DBMS_ CLOUD.PUT_OBJECT procedure:

```
BEGIN
  DBMS_CLOUD.PUT_OBJECT(
  credential_name => 'DEF_CRED_NAME',
  object_uri => 'https://objectstorage.region.oraclecloud.
com/n/namespace-string/b/bucketname/o/',
  directory_name => 'DUMP_DIR'
  file_name => 'Dump file');
END;
/
```

Here, credential_name refers to the Cloud credentials created at the database level. Earlier in this chapter, we discussed the credentials while taking a manual backup of the autonomous database.

Here is an example:

```
Sql>  Begin
DBMS_CLOUD.CREATE_CREDENTIAL(
    credential_name => 'OBJSTORE_CRED',
    username => 'clouduser@gmail.com',
    password => 'iJW61QSOS-4:dHs5CP6B'
  );
END;
```

PL/SQL procedure successfully completed.

The Cloud credential has been created in the database with the name OBJSTORE_CRED. Let's execute the DBMS_CLOUD.PUT_OBJECT procedure to move the dump to object storage.

Let's first move exp01.dmp using this procedure:

```
SQL> BEGIN
DBMS_CLOUD.PUT_OBJECT(
credential_name => 'OBJSTORE_CRED',
object_uri => 'https://swiftobjectstorage.us-ashburn-1.
oraclecloud.com/v1/mytenancy/DBBKP/exp01.dmp',
directory_name => 'DUMP_DIR',
file_name => 'exp01.dmp');
END;
```

PL/SQL procedure successfully completed.

Let's also move exp02.dmp to object storage using this procedure:

```
SQL> BEGIN
DBMS_CLOUD.PUT_OBJECT(
credential_name => 'OBJSTORE_CRED',
object_uri => 'https://swiftobjectstorage.us-ashburn-1.
oraclecloud.com/v1/mytenancy/WalletATP/exp02.dmp',
directory_name => 'DUMP_DIR',
file_name => 'exp02.dmp');
END;
/
```

PL/SQL procedure successfully completed.

We can verify the uploaded objects in **Cloud portal | Object storage**:

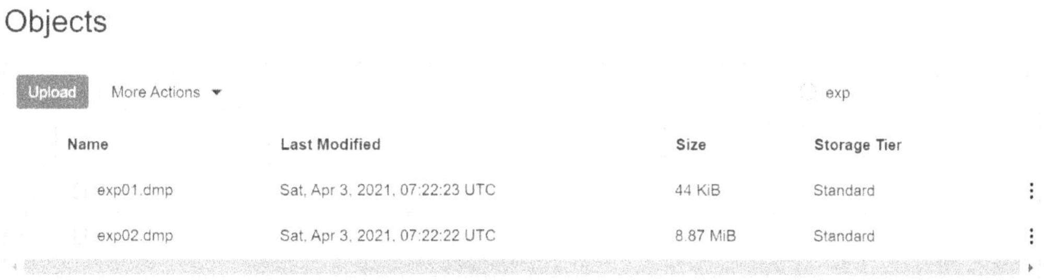

Figure 5.21 – Database backup list

Now let's discuss how to take export directly into object storage instead of creating a dump set.

Creating a Data Pump backup directly to object storage

This procedure is very simple. We need to invoke the expdp command along with the object storage URL:

```
expdp HR/password@ATPDB_high
directory=dump_dir
dumpfile=Credential_name:Object storage URL
parallel=16
encryption_pwd_prompt=yes
```

```
filesize=1G
logfile=export.log
```

If the export utility version is *19.9* or later, then we can specify the credential name as a separate parameter, and the dump file parameter can carry only the object storage URL in this case:

```
expdp HR/password@ATPDB_high
directory=dump_dir
credential=Credential_name
dumpfile=Object storage URL
parallel=16
encryption_pwd_prompt=yes
filesize=1G
logfile=export.log
```

The dump file carries values of the default credential and the object storage URL. We derived values for both of them earlier in this chapter while executing the DBMS_CLOUD.PUT_OBJECT procedure; let's use that.

Here is an example:

```
$ expdp admin@kkatp_high filesize=1GB dumpfile=OBJSTORE_
CRED:https://swiftobjectstorage.us-ashburn-1.oraclecloud.
com/v1/mytenancy/bucket-crsk-ash/o/export%T.dmp parallel=2
encryption_pwd_prompt=yes logfile=export.log directory=data_
pump_dir schemas=test
```

Here is an example of the output:

```
Export: Release 12.2.0.1.0 - Production on Sat Apr 3 11:39:17
2021

Copyright (c) 1982, 2017, Oracle and/or its affiliates.  All
rights reserved.
Password:

Connected to: Oracle Database 19c Enterprise Edition Release
19.0.0.0.0 - Production

Encryption Password:
Starting "ADMIN"."SYS_EXPORT_SCHEMA_01":  admin/********@
```

```
kkatp_high filesize=1GB dumpfile=OBJSTORE_CRED:https://
swiftobjectstorage.us-ashburn-1.oraclecloud.com/v1/ mytenancy
/bucket-crsk-ash/o/export%T.dmp parallel=2 encryption_
pwd_prompt=yes logfile=export.log directory=data_pump_dir
schemas=test
Processing object type SCHEMA_EXPORT/TABLE/TABLE_DATA
Processing object type SCHEMA_EXPORT/TABLE/STATISTICS/TABLE_
STATISTICS
. . exported "TEST"."TEST"                          8.568
MB 1000000 rows
Processing object type SCHEMA_EXPORT/STATISTICS/MARKER
Processing object type SCHEMA_EXPORT/USER
Processing object type SCHEMA_EXPORT/DEFAULT_ROLE
Processing object type SCHEMA_EXPORT/TABLESPACE_QUOTA
Processing object type SCHEMA_EXPORT/PASSWORD_HISTORY
Processing object type SCHEMA_EXPORT/PRE_SCHEMA/PROCACT_SCHEMA
Processing object type SCHEMA_EXPORT/TABLE/TABLE
Processing object type SCHEMA_EXPORT/POST_SCHEMA/PROCACT_SCHEMA
Master table "ADMIN"."SYS_EXPORT_SCHEMA_01" successfully
loaded/unloaded
*****************************************************************
***************
Dump file set for ADMIN.SYS_EXPORT_SCHEMA_01 is:
  https://swiftobjectstorage.us-ashburn-1.oraclecloud.com/v1/
mytenancy /bucket-crsk-ash/o/export20210403.dmp
Job "ADMIN"."SYS_EXPORT_SCHEMA_01" successfully completed at
Sat Apr 3 11:39:58 2021 elapsed 0 00:00:31
```

We can see the export dumps are directly placed in the object storage. We don't need to manually execute the DBMS_CLOUD.PUT_OBJECT procedure.

Data Pump import backup

Now that the dump files are ready, let's talk about the import activity. Import command parameters to autonomous databases are similar to export. Along with normal import parameters, we need to specify the dump file stored in object storage with its URL:

```
$ impdp admin@kkatp_high directory=data_pump_dir
dumpfile=OBJSTORE_CRED:https://swiftobjectstorage.us-ashburn-1.
oraclecloud.com/v1/mytenancy/bucket-crsk-ash/o/export20210403.
dmp  REMAP_SCHEMA=TEST:SCOTT
```

An example of import output is as follows:

```
sImport: Release 12.2.0.1.0 - Production on Sat Apr 3 11:45:20
2021

Copyright (c) 1982, 2017, Oracle and/or its affiliates.  All
rights reserved.
Password:

Connected to: Oracle Database 19c Enterprise Edition Release
19.0.0.0.0 - Production
Master table "ADMIN"."SYS_IMPORT_FULL_01" successfully loaded/
unloaded
Starting "ADMIN"."SYS_IMPORT_FULL_01":  admin/********@kkatp_
high directory=data_pump_dir dumpfile=OBJSTORE_CRED:https://
swiftobjectstorage.us-ashburn-1.oraclecloud.com/v1/ mytenancy /
bucket-crsk-ash/o/export20210403.dmp REMAP_SCHEMA=TEST:SCOTT
Processing object type SCHEMA_EXPORT/USER
Processing object type SCHEMA_EXPORT/DEFAULT_ROLE
Processing object type SCHEMA_EXPORT/TABLESPACE_QUOTA
Processing object type SCHEMA_EXPORT/PASSWORD_HISTORY
Processing object type SCHEMA_EXPORT/PRE_SCHEMA/PROCACT_SCHEMA
Processing object type SCHEMA_EXPORT/TABLE/TABLE
Processing object type SCHEMA_EXPORT/TABLE/TABLE_DATA
. . imported "SCOTT"."TEST"                              8.568
MB 1000000 rows
Processing object type SCHEMA_EXPORT/TABLE/STATISTICS/TABLE_
STATISTICS
Processing object type SCHEMA_EXPORT/STATISTICS/MARKER
Processing object type SCHEMA_EXPORT/POST_SCHEMA/PROCACT_SCHEMA
Job "ADMIN"."SYS_IMPORT_FULL_01" successfully completed at Sat
Apr 3 11:45:40 2021 elapsed 0 00:00:10
```

Please note that Autonomous Database supports external tables, partitioned external tables, and external partitions of hybrid partitioned tables. But the backup will not include those external objects. It has to be backed up separately through different methods.

Refreshing Autonomous Database schemas

In real-time scenarios, we may have a requirement to refresh user data, such as the Dev database may need to be refreshed periodically from production. This requirement could appear in an autonomous environment as well. Imagine that the Dev and Production databases are autonomous databases and we may need to refresh Dev data from the Production database, or we may need to refresh Autonomous Database schema from an on-premises database, or vice-versa. Let's discuss how we can achieve this.

In general, we can do a data refresh through the expdp/impdp procedure that we discussed earlier in *Manual database backup using Data Pump*. We can export the required data from the source database and import it into Autonomous Database.

Another method is through a database link. Create a database link between these databases and pull or push the data according to the need. We are going to discuss how to create database links in Autonomous Database.

Database links between autonomous databases

In general, a database link is used to communicate between different databases. The beauty of this method is that we don't need to worry about compatibility between the databases. The source and target database could be with different platforms, different versions, and different character sets. The only requirement is SQL connectivity between these databases. We create a link between schemas present in source and target databases. From the source database, we can refer data of schema located in the target database.

Autonomous Database supports creating database links only if the target database is accessible through a public IP or public hostname or using an Oracle Database Gateway.

In Autonomous Database, the database link can be created using the DBMS_CLOUD_ADMIN. CREATE_DATABASE_LINK procedure:

```
DBMS_CLOUD_ADMIN.CREATE_DATABASE_LINK(
        db_link_name          IN VARCHAR2,
        hostname              IN VARCHAR2,
        port                  IN NUMBER,
        service_name          IN VARCHAR2,
        ssl_server_cert_dn    IN VARCHAR2,
        credential_name       IN VARCHAR2,
        directory_name        IN VARCHAR2 DEFAULT,
        gateway_link          IN BOOLEAN DEFAULT);
```

Here, the parameters and their definitions are as follows:

- db_link_name: Database link name
- Hostname: Target database hostname
- Port: Port at which the target database listener is active

- Service_name: Target database service name
- ssl_server_cert_dn: The DN value found in the server certificate
- credential_name: Credential to log in to the target autonomous database

This is different from the credential that we created earlier for expdp. For the expdp command, the credential parameter had Cloud portal login credentials. But here, we create credentials for the database, and hence the database schema should be specified.

Here is an example:

```
BEGIN
DBMS_CLOUD.CREATE_CREDENTIAL(
credential_name => 'DB_Schema_credential',
username => 'ADMIN',
password => 'Welcome123##'
);
END;
```

The attributes and their descriptions are as follows:

- directory_name: Directory in which the cwallet.sso wallet has been stored. This directory location will be within an autonomous environment. As per our previous example, it is the /u03/dbfs/BDF28FA2296E84A6E053C614000ACEA2/data folder.
- gateway_link: Indicates whether the target is an autonomous database/any other database or Database Gateway.

In a general autonomous database, the authentication happens through the wallet. To use database links with Autonomous Database, the target database must be configured to use TCP/IP with SSL (TCPS) authentication. So, let's first download the wallet. We can retrieve the hostname, port, and service name values from the wallet. The steps are as follows:

1. Log in to the OCI portal, and navigate to **Autonomous Database**. Choose **ATP** or **ADB** according to the workload:

Figure 5.22 – ATP DB connection

2. Click on the **DB Connection** button, as shown. It will invoke the **Database Connection** page:

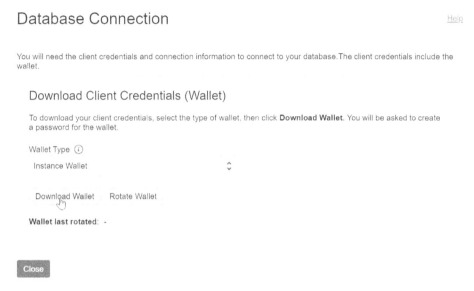

Figure 5.23 – Wallet download

3. Click on the **Download Wallet** button, which will ask for a password to protect the wallet download:

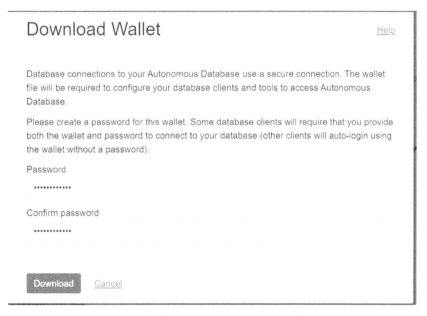

Figure 5.24 – Wallet download credentials

The DB connection will be downloaded as a ZIP file. Extract the ZIP file and you can find the `cwallet.so` wallet, `tnsnames.ora`, and other network-related files. The wallet is available now, but it should be on the Autonomous Database server. To transfer the downloaded wallet to the Autonomous Database server, first, we will move it to object storage and then move it to the Autonomous Database server by executing the `DBMS_CLOUD.GET_OBJECT` procedure. Transferring the wallet to object storage can be done through the Cloud portal or using the OCI **command-line utility (CLI)**.

After copying/moving the wallet to object storage, execute `DBMS_CLOUD.GET_OBJECT` as shown to copy it to the Autonomous Database server:

```
BEGIN
DBMS_CLOUD.GET_OBJECT(
credential_name => 'DEF_CRED_NAME',
object_uri => 'https://objectstorage.us-phoenix-1.oraclecloud.
com/n/<tenancyName/Namespace>/b/<bucket_name>/o/cwallet.sso',
directory_name => 'DATA_PUMP_DIR');
END;
```

Here, the attributes and descriptions are as follows:

- `Credential name`: The Cloud credentials created at the database level
- `Object_uri`: Object storage URL after uploading the wallet to object storage
- `Directory_name`: Directory in which the wallet will be transferred, basically the logical directory created at the autonomous database level

Here is an example:

```
BEGIN
    DBMS_CLOUD.GET_OBJECT(
    credential_name => 'OBJSTORE_CRED',
    object_uri => 'https://objectstorage.usphoenix-1.
oraclecloud.com/n/mynamespace/b/bucketname/o/cwallet.sso',
    directory_name => 'DUMP_DIR');
END;
/
```

The `https://objectstorage.usphoenix-1.oraclecloud.com/n/mynamespace/b/bucketname/o/cwallet.sso` wallet has been copied to the `DUMP_DIR` logical directory. Now, we have everything to create a database link.

Let's revisit the database link procedure with collected attributes:

```
DBMS_CLOUD_ADMIN.                          DBMS_CLOUD_ADMIN.
CREATE_DATABASE_LINK                       CREATE_DATABASE_LINK
(                                          (
db_link_name      IN VARCHAR2,             DBlink_autonomous,
hostname          IN VARCHAR2,             adb.<region>.oraclecloud.com,
port              IN NUMBER,               1522,
service_name      IN VARCHAR2,             lstssxuegqjldw2_<DB_name>.adb.oraclecloud.com
ssl_server_cert_dn IN VARCHAR2,            "CN=adwc.uscom-east.oraclecloud.com,OU=Oracle
                                           BMCS US,O=Oracle Corporation,L=Redwood
                                           City,ST=California,C=US"
credential_name   IN VARCHAR2,             'DB_Schema_credential',
directory_name    IN VARCHAR2,             'DUMP_DIR',
gateway_link      IN BOOLEAN               False
);                                         );
```

Figure 5.25 – Database link procedure with attributes

Execute this procedure from the source database to create a database link to the target database. Once the link is created between schemas of two different databases, we can write SQL scripts to perform a refresh of a particular table or schema or the whole database itself.

Summary

In this chapter, we discussed the backup/restore concepts of Autonomous Database. We learned how automated backup works with Autonomous Database and then discussed the available manual backup methods. We also discussed how the backup restoration process is automated in Autonomous Database. This chapter also covered the database link mechanism, which plays a crucial role in database refresh activity.

In the next chapter, we will be discussing how high availability can be achieved for Autonomous Database through Oracle Data Guard.

Questions

1. Can we do point-in-time recovery in Autonomous Database?

2. Is it possible to refresh a particular table between different autonomous databases?

Further reading

- *Oracle Database 12c Backup and Recovery Survival Guide*, by Francisco Munoz Alvarez and Aman Sharma

Answers

1. Yes. We can do point-in-time recovery in Autonomous Database. This option is available in the Autonomous Database portal.

2. Yes. We can refresh tables between different autonomous databases through the database link.

6

Managing Autonomous Databases

In earlier chapters, we discussed how autonomous databases are provisioned and the available options, the migration strategies, the disaster recovery concepts, and the backup/recovery options. In this chapter, we are going to explore the manageability of autonomous databases.

In general, a database is effective when it is managed well. This management topic covers how to operate and manage an autonomous database whether it is an **Autonomous Transaction Processing (ATP) database**, an **Autonomous Data Warehouse** (**ADB**), an APEX database (built for Oracle APEX application development), or a JSON (an autonomous database for JSON-centric applications) database.

This chapter will explore all the management options and functionalities of autonomous databases. We will discuss which tasks are automated and which management tasks need to be managed by us. At the end of this chapter, we will have a better understanding of how to effectively manage an autonomous database.

This chapter also covers the basic day-to-day tasks of a database administrator, such as the following:

- How to start/stop the database
- Scaling up/down
- Connecting to an autonomous database
- Monitoring the performance of the database
- Cloning the database

Let's get started.

Technical requirements

To follow the instructions given in this chapter, you'll need an **Oracle Cloud Infrastructure** (**OCI**) free tier/pay-as-you-go autonomous database (ATP or ADB).

Autonomous database actions

Let's start by exploring all the options on the autonomous database **Overview** page. Log in to the OCI portal using your own credentials, then click the **Oracle Database** option, and then **Autonomous Database**, as shown in *Figure 6.1*. Choose either **Autonomous Transaction Processing**, **Autonomous Data Warehouse**, or **Autonomous JSON Database**.

Figure 6.1 – Oracle Cloud database options

The autonomous database details are shown in *Figure 6.2*. This details page shows information such as **Database Name**, **OCPU Count**, and **Storage**, and it also has options to manage the database, starting with **Start** or **Stop** (depending on the database's current status), and options for managing the database. We will be exploring each option in detail in this chapter.

Figure 6.2 – Autonomous database overview

Starting or stopping an autonomous database

The first important topic to discuss is how to start and stop the autonomous database. The start and stop actions can be performed via the OCI portal, the **OCI Command-Line Interface** (**OCI CLI**), or a REST API:

- **OCI portal**: As shown in *Figure 6.2*, on the autonomous database overview page we can find the **More Actions** button. If we click on that, we will see many options, such as **Start** or **Stop** (depending on the current database status), as shown in *Figure 6.3*. We can also see the database workload type, which is ATP. If the database is stopped, then the letters ATP will have an orange background. Click the **Start** button to start the database.

Figure 6.3 – Start the autonomous database

Once the database has been started, the ATP background changes to green. The database status will change to available, as shown in *Figure 6.4*:

Figure 6.4 – Autonomous database details

- **OCICLI**: Autonomous databases can be started and stopped via the OCI CLI with the following commands:

```
oci db autonomous-database start –autonomous-database-id <OCID
of Autonomous database>
oci db autonomous-database stop –autonomous-database-id <OCID
of Autonomous database>
```

We need to pass the OCID of the autonomous database as an argument in OCICLI to start and stop the database. OCID information can be collected from the overview page.

- **REST API**: One way to start or stop an autonomous database is via a REST API:

  - **API to start the database**: POST /20160918/autonomousDatabases/ {autonomousDatabaseId}/actions/start

  - **API to stop the database**: POST /20160918/autonomousDatabases/ {autonomousDatabaseId}/actions/stop

There are many tools, such as Terraform and oci-curl, that you can use to invoke a REST API function. We can start and stop the database using the API without interacting with the OCI portal. Also, APIs are very useful when we want to automate this functionality through scripts.

Now that the autonomous database has been started, the next step is to make a connection with it. Let's discuss that in the next section.

Autonomous database connection

Autonomous database connectivity is available as either public (available to all) or private (accessible within the same **virtual cloud network (VCN)**). We will discuss each connection type in detail in the upcoming sections. The autonomous database connection is always encrypted, so we need a wallet file to connect with the database. Click **DB Connection** on the **Autonomous Database Information** page. An autonomous database can be created with either a Shared Infrastructure or a Dedicated Infrastructure. If it is a Shared Infrastructure, then clicking the **DB Connection** button will invoke another window, as shown in *Figure 6.5*.

The autonomous database has two wallet types – **Instance Wallet** and **Regional Wallet**. You can find them in the **Wallet Types** drop-down box. **Instance Wallet** provides keys and wallets that are required to connect to this single autonomous database. **Regional Wallet** provides the keys and wallets required to connect all autonomous databases in that region.

In general, Regional Wallet is used by database administrators or account administrators.

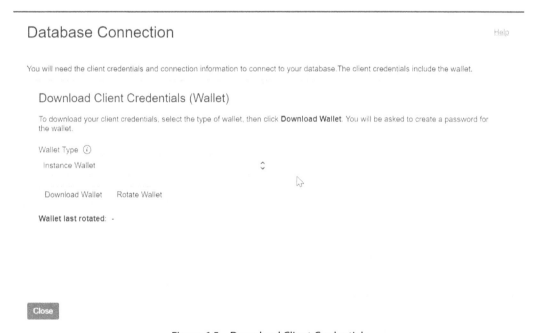

Figure 6.5 – Download Client Credentials

As we can see in *Figure 6.5*, there are two options for downloading client credentials – you can either **Download Wallet** or **Rotate Wallet**. We will discuss the **Rotate Wallet** option in the next section.

To download an **Instance Wallet** or a **Regional Wallet type**, we need to provide a password, as shown in *Figure 6.6*. This password is required to avoid unauthorized access. Also, database tools that do not have auto-login will require the wallet and password to make connections.

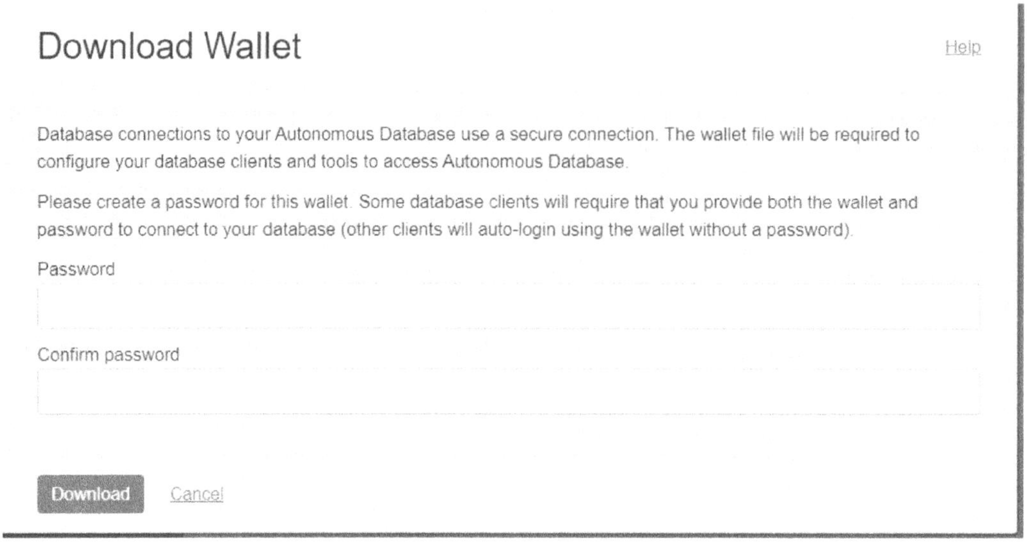

Download Wallet   Help

Database connections to your Autonomous Database use a secure connection. The wallet file will be required to configure your database clients and tools to access Autonomous Database.

Please create a password for this wallet. Some database clients will require that you provide both the wallet and password to connect to your database (other clients will auto-login using the wallet without a password).

Password

Confirm password

[Download]   Cancel

Figure 6.6 – Download Wallet

After providing a password, click the **Download** button. It will download the wallet as a `.zip` file (`Wallet_<ORACLE_SID>.zip`) in the default download location. The `.zip` file will contain all the files required to make SQL *Plus connections with the autonomous database. In general, for SQL Plus connectivity, we need `sqlnet.ora`, `tnsnames.ora`, and a wallet file as the database connection is encrypted in an autonomous environment.

The wallet ZIP file contains the following files:

- `Readme`: This contains details about the wallet, such as its expiry date. In general, a wallet is valid for 21 months. However, it can be downloaded again at any time to extend its validity. This folder also has autonomous database tools and resources with access links. With these access links, we can directly access the tools via the browser without logging in to the OCI portal and exploring the autonomous page.

- `cwallet.sso`: This is the Oracle auto-login wallet used for authorization.

- `ewallet.p12`: This is the wallet file associated with the auto-login wallet.

- `sqlnet.ora`: This is a SQL*Net profile configuration file that includes the wallet location and the TNSNAMES naming method.

- `tnsnames.ora`: This is a SQL*Net configuration file that contains network service names mapped to connect descriptors for the local naming method. Autonomous databases provide five different connection descriptors (HIGH, LOW, MEDIUM, TP, AND TPURGENT) to handle different kinds of workload. Each connection descriptor has a different priority with different resource allocations.

- `keystore.jks`: This is a **Java KeyStore (JKS)** file for JDBC Thin connections.

- `ojdbc.properties`: This carries the wallet location. The JDBC driver retrieves the username and password from the wallet.

- `truststore.jks`: Truststore is used to store certificates from **Certificate Authorities (CAs)** that verify the certificate presented by the server in SSL connections.

Basically, the unzipped folder has all the files required to make a SQL connection with the autonomous database. In the SQL client machine, the `TNS_ADMIN` environment variable should point to the unzipped folder (created by extracting this `.zip` file). We should make sure that the wallet is only available to authorized people.

After setting the `TNS_ADMIN` variable, we can now make a connection to an autonomous database. By default, SQL Plus connectivity with autonomous databases requires the ADMIN user. We provided the password for the ADMIN user when we provisioned the database. With the wallet and the ADMIN user password, we can now log in to the autonomous database.

You can reset the password of the Admin user through the portal. We will discuss this in the *Administrator password* section.

As a security best practice, we should rotate the wallet at regular intervals because using the same wallet for a long period is a security threat. After rotation, we will get a new set of certificate keys and credentials. Also, rotating an instance or regional wallet requires confirmation, as shown in *Figure 6.7*:

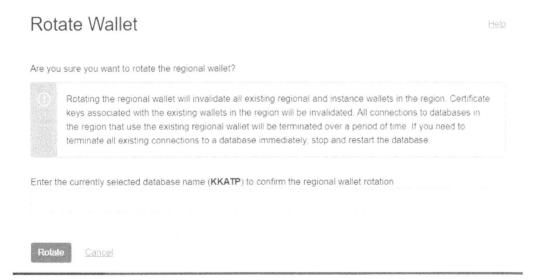

Figure 6.7 – Rotate Wallet

There are a few changes after wallet rotation. Old keys and credentials will become invalid after rotation. You need to download the wallet again to make successful connections, and existing user sessions will be terminated. As we know, there are two kinds of wallet: Instance Wallet and Regional Wallet. Rotating the Instance Wallet type will not invalidate the Regional Wallet type that covers the same database instance, whereas rotating the Regional Wallet type will invalidate the keys of all databases in that region. We will get a detailed message about the rotation's progress, as shown in *Figure 6.8*:

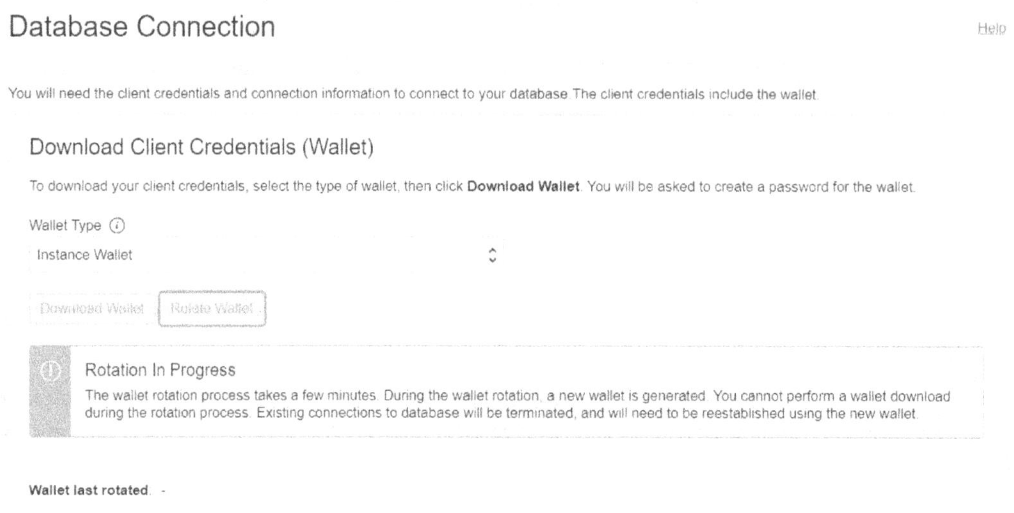

Database Connection   Help

You will need the client credentials and connection information to connect to your database. The client credentials include the wallet.

Download Client Credentials (Wallet)

To download your client credentials, select the type of wallet, then click **Download Wallet**. You will be asked to create a password for the wallet.

Wallet Type (i)

Instance Wallet

[Download Wallet] [Rotate Wallet]

(i) Rotation In Progress
     The wallet rotation process takes a few minutes. During the wallet rotation, a new wallet is generated. You cannot perform a wallet download during the rotation process. Existing connections to database will be terminated, and will need to be reestablished using the new wallet.

Wallet last rotated -

Figure 6.8 – Wallet rotation progress

We can see information about when this wallet has been rotated before. The next topic that we are going to discuss is **Performance Hub**.

Performance Hub

The next important management feature in the autonomous database is **Performance Hub**. As we know, an autonomous database manages its performance on its own. At the same time, we can see insights into the autonomous database's real-time CPU usage, memory usage, user I/O, and disk I/O.

To access the Performance Hub page, click on the **Performance Hub** button on the **Autonomous Database Information** page.

At the top of the page, we can find the **Activity Summary** window, which displays the average active sessions. It provides a graphical representation of **User I/O**, **Wait**, time, and **CPU** usage. To look for specific data, we could filter time range (last hour/last 8 hours/last week/any custom range). We can also change the time zone, as shown in *Figure 6.9*. Below **Activity Summary** (**Average Active Sessions**), we can find **ASH Analytics**, **SQL Monitoring**, **ADDM**, **Workload**, and **Blocking Sessions**.

Figure 6.9 – Performance Hub – Average Active Sessions

The **Performance Hub** page has many useful details that can help us to analyze the performance of our database. At the top, you can also find the **Reports** button, which can generate an **Automatic Workload Repository (AWR)** report. Click **Automatic Workload Repository** in the drop-down list, as shown in *Figure 6.10*.

Figure 6.10 – Performance Hub – AWR

It will open a new window, as shown in *Figure 6.11*. The new page allows us to choose the snapshot timings to generate the AWR report.

Figure 6.11 – Performance Hub – AWR period

Click **Download** to generate the AWR report and download it to your local system.

We will discuss the performance metrics (**ASH Analytics**, **SQL Monitoring**, **ADDM**, **Workload**, and **Blocking Sessions**) collected by our autonomous database in the following sections.

ASH Analytics

The **Active Session History** (**ASH**) chart explores all active session data through the SQL ID and user sessions. An active session is a session that is actively using the CPU, rather than waiting for a wait event. Using **ASH Analytics**, we can analyze short-lived performance issues from various dimensions, such as SQL ID, time, module, and action, as shown in *Figure 6.12*.

Figure 6.12 – Performance Hub – ASH Analytics

This **ASH Analytics** graph can be seen from various perspectives. Let's examine each section in detail. As shown in *Figure 6.13*, in the **Average Active Sessions** section, we can see **ASH Dimensions**. Click on the drop-down box.

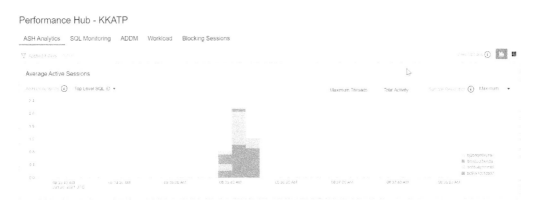

Figure 6.13 – Performance Hub – ASH Analytics (Average Active Sessions)

The drop-down box shows various dimensions, as shown in *Figure 6.14*. By default, **Top Dimensions** will be shown for SQL ID. We also have the option to view this graph from a SQL perspective, which will get us SQL information from every session.

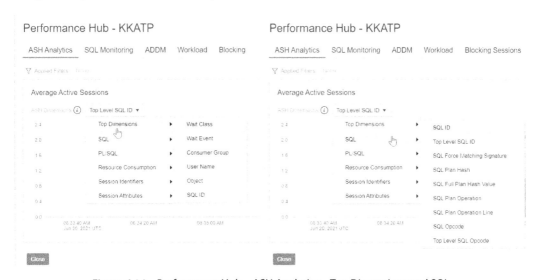

Figure 6.14 – Performance Hub – ASH Analytics – Top Dimensions and SQL

The graph also has the option to view sessions that have PL/SQL executions, as shown in *Figure 6.15*. A separate dimension is available to view **Resource Consumption**, which provides details about wait classes and events.

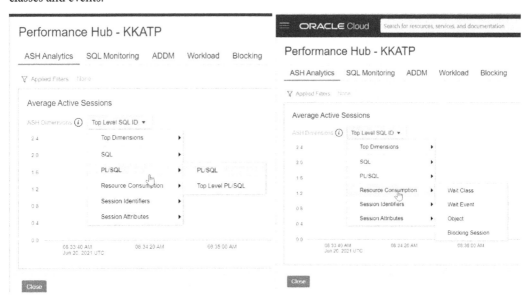

Figure 6.15 – Performance Hub – ASH Analytics – PL/SQL

It also has the option to view PL/SQL-specific operations and each session's attributes, as shown in *Figure 6.16*.

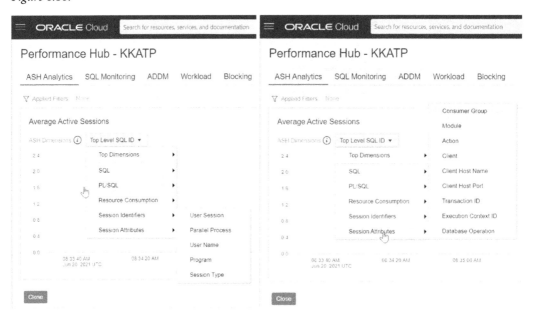

Figure 6.16 – Performance Hub – ASH Analytics – Session Attributes

The graph also has dimensions for **Session Identifiers** and **Session Attributes**, which will project the graph in terms of user session details, which we use to see the **v$session** view in the database.

SQL monitoring

SQL statements that have taken more than 5 seconds to execute will be monitored, and their database time, CPU time, and I/O requests will be shown in the **SQL Monitoring** section of the **Performance Hub** graph, as shown in *Figure 6.17*. This information will be useful when we have long-running SQL queries.

Figure 6.17 – Performance Hub – SQL Monitoring

ADDM

Automatic Database Diagnostic Monitor (**ADDM**) is a performance monitoring tool that examines the data collected by the AWR on a regular basis and if there are any performance problems, it identifies the root cause of these problems and provides recommendations to correct them. As shown in *Figure 6.18*, the findings of ADDM will be listed with recommendations.

Figure 6.18 – Performance Hub – ADDM

Workload

The **Workload** tab displays four sets of database statistics: **CPU Statistics**, **Wait Time Statistics**, **Workload Profile**, and **Sessions**.

The **CPU Statistics** chart has details about CPU time and CPU utilization. This chart shows CPU time consumed by foreground sessions, any unusual CPU spikes, and the percentage of CPU time used by each consumer group.

The **Wait Time Statistics** chart displays the time used in different events.

The **Workload Profile** has a group of charts indicating patterns of user calls, transactions, and executions.

The **Sessions** chart shows the number of current successful logins and the number of sessions.

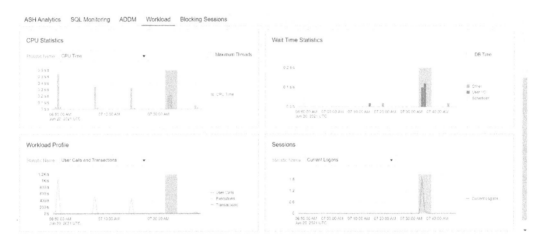

Figure 6.19 – Performance Hub – Workload

Blocking Sessions

The **Blocking Sessions** chart contains detailed information about the blocking sessions, as shown in *Figure 6.20*. The chart also gives information about the user sessions that are blocked by these blocking sessions. It also provides SQL involved in blocking session, which helps to find the root cause of the blockage. The page also allows us to kill the blocking sessions.

Figure 6.20 – Performance Hub – Blocking Sessions

With this, we have covered **Performance Hub**. The next section to explore is **Scale Up/Down**.

Scale Up/Down

By default, the autonomous database will be created with 1 OCPU and 1 TB of storage. However, it can be scaled up or down either manually or automatically. On the **Autonomous Database Information** page, click the **Scale Up/Down** button.

Figure 6.21 – Scale Up/Down

It will take you to the **Scale Up/Down** window, as shown in *Figure 6.22*.

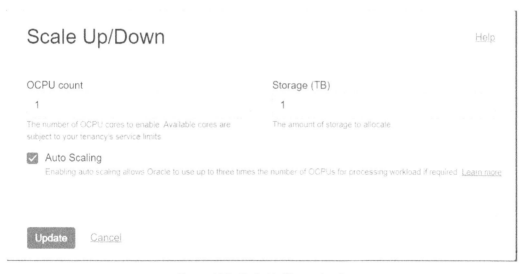

Figure 6.22 –Scale Up/Down details

We can choose the required OCPU count and the amount of storage in TB. We also have the option to specify **Auto Scaling**. CPU scaling doesn't require downtime, and the billing will be calculated on average OCPU usage per hour.

Auto Scaling is enabled by default, and it will scale up three times the current CPU base number. If the base number is 1 OCPU, then auto scaling will scale up the OCPU count to 3, which continues as demand increases. In the same way, scaling down will happen when demand decreases. In a Shared Infrastructure environment, where multiple autonomous databases share the same infrastructure, auto scaling happens on a first-come, first-served basis. In a Dedicated Infrastructure environment, auto scaling will happen until the maximum available number of cores for that Exadata infrastructure minus the number of OCPUs consumed by other databases in the same infrastructure. For example, for an Exadata X8 quarter rack Dedicated Infrastructure, the maximum number of OCPUs is 100. If more than one database is deployed in that environment, then auto scaling will calculate the available OCPUs as 100 minus the number of OCPUs used by all databases.

Auto scaling or storage will happen at any time without downtime and performance impact.

Let's discuss some of the other options on the **Autonomous Database Information** page in the **More Actions** dropdown, as shown in *Figure 6.23*.

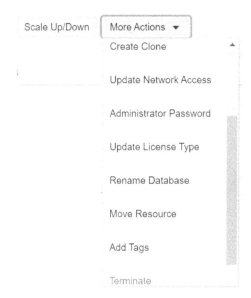

Figure 6.23 – More Actions of autonomous databases

Cloning

Database cloning is always interesting. We may need to clone a database for testing or recovery purposes. Autonomous databases also have the option to be cloned. The autonomous database clone can be created using the **Create Clone** option, as shown in *Figure 6.23*. Cloning is fully automated here.

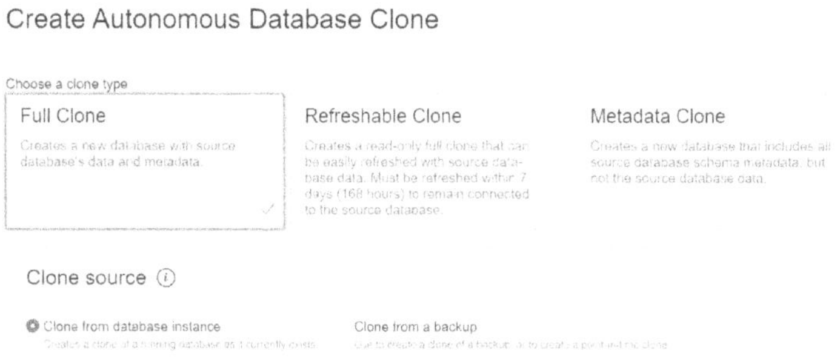

Create Autonomous Database Clone

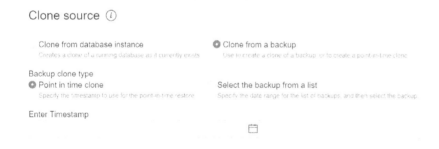

Figure 6.24 – Create Autonomous Database Clone

The clone can be created in three ways, as we can see in *Figure 6.24*:

- **Full Clone**: The cloned database will have all the data and metadata of the source database.

- **Refreshable Clone**: This is a read-only clone of the source database.

- **Metadata Clone**: This is a clone of the source database without data.

For **Full Clone** and **Metadata Clone**, we can choose to **Clone from database instance** or **Clone from a backup**.

Figure 6.25 – Clone source

For the **Backup clone type** option, we can choose a particular timestamp or a backup from the list, as shown in the screenshot. The backup should be at least 2 hours old.

In the next step, basic details such as compartment, **Source database name**, and **Database name** should be chosen.

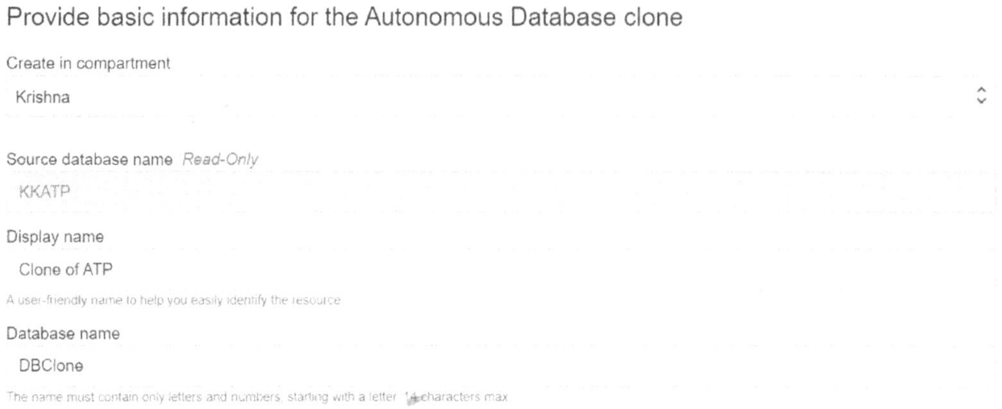

Provide basic information for the Autonomous Database clone
Create in compartment
Krishna
Source database name Read-Only
KKATP
Display name
Clone of ATP
A user-friendly name to help you easily identify the resource
Database name
DBClone
The name must contain only letters and numbers, starting with a letter. 14 characters max.

Figure 6.26 – Clone database creation details

Provide configuration details for the cloned database, as shown in *Figure 6.27*.

Configure the database
Choose database version
19c
OCPU count
1
Storage (TB)
1
Auto scaling

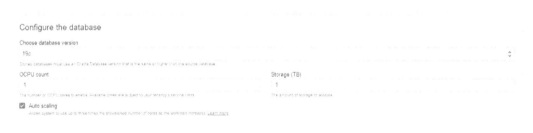

Figure 6.27 – Clone database configuration

We need to choose the database version, which will be always the latest version. **OCPU count** and the **Storage (TB)** size should be provided for the clone database. The OCPU count can be specified up to 128. Enabling **Auto scaling** helps to manage the resource need automatically. As we discussed in the previous section, when there is a demand, it scales up the OCPU of the autonomous database instance up to three times the assigned OCPU value, and it doesn't require downtime. Also, it automatically reduces the number of cores when demand is lower. The storage limit will be specified in **terabytes (TB)** and we can specify up to 128 TB.

Then provide Admin credentials for the cloned database. The password should have a minimum of one uppercase character, one lowercase character, and one number; it should not be one of the last three passwords used; and it should also not contain the word admin. It can contain special characters, but double quotation marks are not allowed. For a refreshable clone, the Admin credentials should be the same as the source database credentials.

Create administrator credentials (i)

Username *Read-Only*

   ADMIN

ADMIN username cannot be edited

Password

Confirm password

Figure 6.28 – Administrator credentials for the cloned database

For the cloned database, we need to define the network access and the license type, as shown in the following screenshot. It could be different from the source database network settings. Network access defines how the cloned database should be accessed. The access happens via SQLNet connectivity. This network access mechanism defines whether public access is allowed for SQLNet connectivity or is possible only within the VCN where the cloned database is located.

Figure 6.29 – Cloned database network access

The **Secure access from everywhere** option allows all open connections to the database, but when access control rules are configured, then it provides the choice to restrict the access to a set of IP addresses or any CIDR block, VCN, or VCN's OCID, as shown in *Figure 6.30*.

Figure 6.30 – Cloned database network access

Private endpoint access only needs VCN and subnet details to allow communication from a particular network of the same tenancy. Also, a network security group can be specified, as shown in the following screenshot. We can modify the network access method after the autonomous database is created. We will discuss that in the next section.

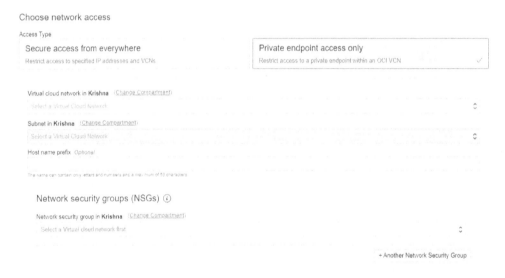

Figure 6.31 – Cloned database – private endpoint access

The license could be **Bring Your Own License** (**BYOL**) or let the usage include database licensing fees as well. BYOL allows users to use the licenses that they currently own for on-premises software. We could also provide up to 10 maintenance contacts (email IDs) for this cloned database. Click the **Create Autonomous Database Clone** button to initiate the action.

Updated Network Access

The network access settings define how the autonomous database should be accessed. We can modify the network access method after the autonomous database has been created as well. Click the **Update Network Access** option available under the **More Actions** button in the details page.

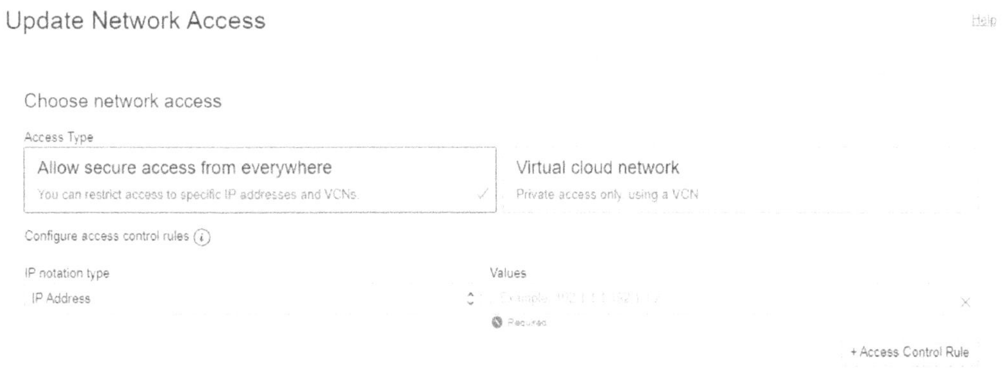

Figure 6.32 – Clone database – Access from everywhere

Secure access from everywhere comes with access control rules by default. We can specify IP addresses, CIDR values, any VCN in the same tenancy, or a VCN's OCID. When we choose a VCN, all subnets belonging to this VCN will get connectivity to our database.

We can only get communication within a VCN in the same tenancy. We need to specify the VCN and the subnet to which we would like this database to connect. With this option, we give access to a specific subnet where it differs from the **Secure access from everywhere** method. Also it asks to specify **Network Security Group** (**NSG**) to adhere to the rules imposed on this environment.

Access control list

The **access control list** protects the autonomous database by filtering IP addresses and VCNs. Only the allowed IP addresses and VCNs can make a connection with an autonomous database. ACL is applicable only to autonomous databases with a network type of **Allow secure access from everywhere**. An access control list should contain a minimum of one entry of an IP address or a range of addresses.

Access control lists can be enabled on the **Autonomous Database Information** page. The networking section of the details page contains the option to **Enable** or **Edit** an access control list.

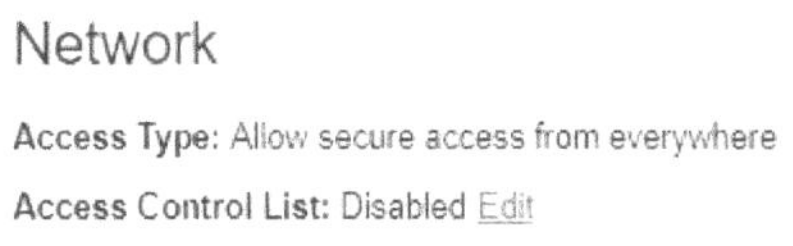

Figure 6.33 – Enabling an access control list

To edit an access control list, click the **Edit** button as shown in *Figure 6.33*. A new window will appear.

Figure 6.34 – Autonomous database – Edit Access Control List

An access control list can be created using either an IP address, a CIDR block, or a VCN. After providing the necessary values, save the list, and that will enable the access control list for the database.

Administrator password

The administrator in an autonomous database equates to the Admin user located in the pluggable database. At the time the autonomous database was created, we specified the password. After the database has been created, it can be reset or changed at any time using the **Administrator** option.

The password needs to follow certain standards. It should contain a minimum of 12 characters and a maximum of 30 characters with at least one lowercase character, one uppercase character, and one number. Also, it should not contain certain special characters, such as double quotation marks, or "admin," regardless of casing. The password policy considers the history of passwords, so it should not be one of the last four passwords used for the database, and the same password should not have been used within the last 24 hours.

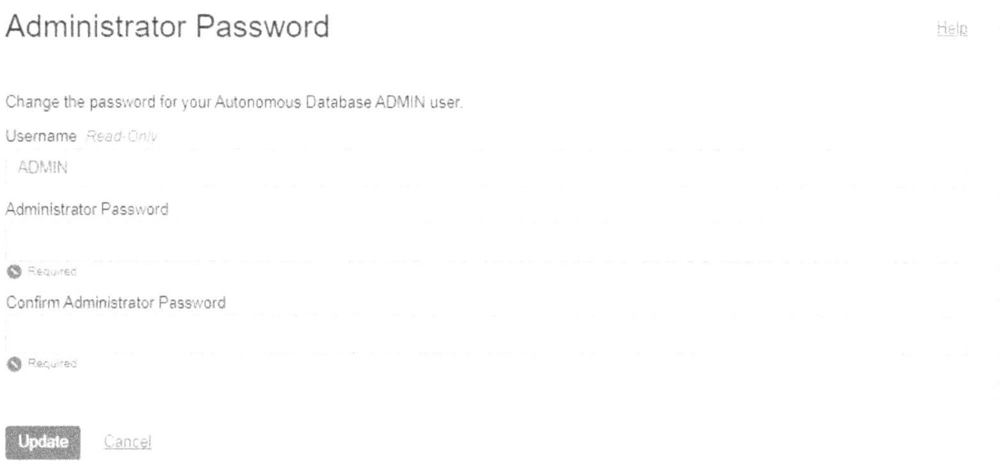

Administrator Password

Change the password for your Autonomous Database ADMIN user.

Username *Read-Only*

ADMIN

Administrator Password

❌ Required

Confirm Administrator Password

❌ Required

Update   Cancel

Figure 6.35 – Autonomous database Administrator Password

The next important thing to discuss is updating the license type.

Updating the license type

Autonomous database pricing can be either unit pricing or BYOL. The unit pricing model will charge for what we use; it's a pay-as-you-go model. BYOL allows us to leverage existing on-premises licenses to move to Oracle Cloud with equivalent, highly automated Oracle PaaS and IaaS services in the cloud.

Update License Type allows us to switch the license option at any time, as shown in *Figure 6.36*. Charging will be calculated based on the option chosen.

Figure 6.36 – Autonomous database license type

Renaming a database

The autonomous database can be renamed at any time using this option. The new name should have only letters and numbers, with a maximum of 14 characters. Special characters are not allowed. Note that the autonomous database OCID will not change during this operation, but the database will be restarted.

Figure 6.37 – Renaming the autonomous database

The rename operation requires a few subsequent activities because some of the operations depend on the database name and connection string:

- The wallet will be modified after a rename operation, hence it needs to be downloaded subsequently. This is applicable to Regional Wallet types as well.

- The connection string referred to at **Application** or **Tools** should be modified according to the new connection string.

- If a database link has been created for this database, then the link should be recreated using the new connection string.

- The rename operation cannot be done to databases that are involved in Data Guard or refreshable clone activities.

The next commonly used option is moving resources, that is, moving an autonomous database to a different compartment.

Move Resource

The **Move Resource** option is used to move an autonomous database to a different compartment in the same tenancy.

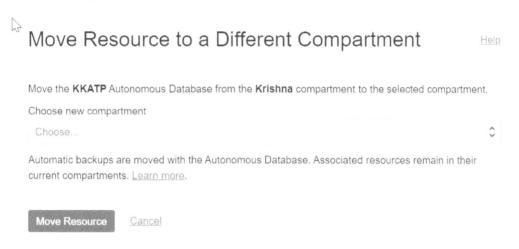

Figure 6.38 – Moving a database to a different compartment

Choose the new compartment, as shown in *Figure 6.38*, and click the **Move Resource** button. The autonomous database will be mapped to a new compartment. No changes are required after this operation.

The service console

Though an autonomous database is a managed environment, there are common questions, such as "How we can monitor our autonomous environment? How can we make sure that CPU and memory are effectively utilized by our autonomous database?" Basically, this information is very much required when we think about scaling up the resources.

We are going to talk about autonomous database monitoring tools and resources in detail in this section.

Let's begin with the Service Console, which has tools for monitoring, administration, and development. The **Service Console** button is on the **Overview** page of our autonomous database, as shown in *Figure 6.39*.

Figure 6.39 – Autonomous database Service Console

Click the **Service Console** button. It will open a new window asking for database credentials, as shown in *Figure 6.40*. The database credentials could be for the Admin user or for any other database user. According to the user privileges, database metrics will be collected.

Figure 6.40 – Credentials for the Service Console

Once the database user credentials are entered, click on the **Sign in** button, as shown in *Figure 6.40*. It will direct you to the autonomous database **Overview** page:

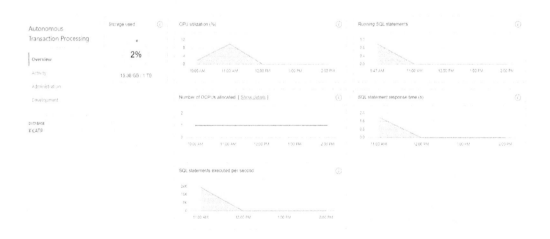

Figure 6.41 – Autonomous database – Service Console – Overview page

The Overview page has details about how much storage and how much of the CPU is consumed by this database. It also has information about the number of OCPUs allocated to this database, SQL statement execution, and response time.

On the overview page, we can see **Activity**, **Administration**, and **Development** options. The **Administration** and **Development** options provide tools to perform administration and development jobs. The **Activity** page shows details about database activity and executed SQL queries, and displays the details in graphs, as shown in *Figure 6.42*.

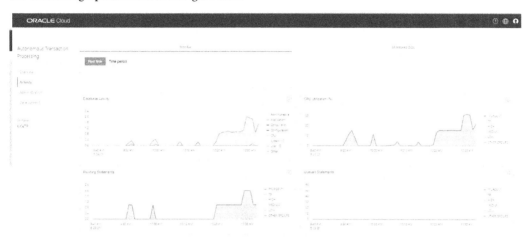

Figure 6.42 – Autonomous database – Service Console – Activity

On the same page, the **Monitored SQL** tab shows details of the SQL statements.

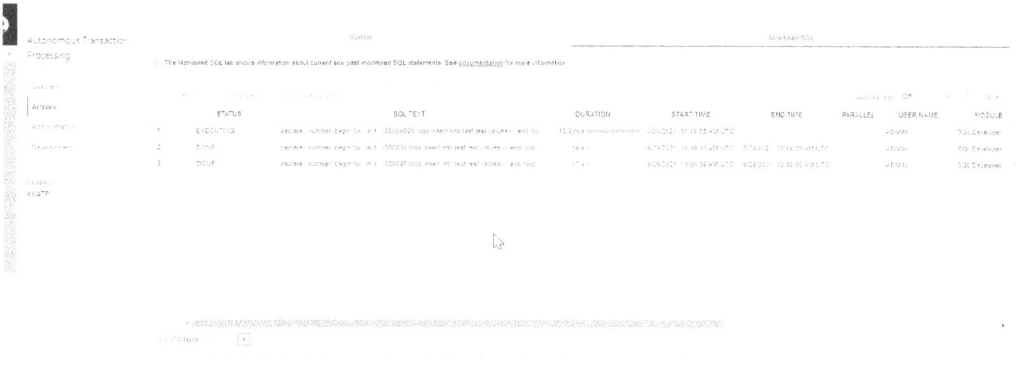

Figure 6.43 – Autonomous database – Service Console – Activity

We can see details about SQL commands, the duration and status of execution, the username, and the module executing the SQL commands.

Changing the workload type

We specify the workload type at the time the autonomous database was created. We will have to choose the appropriate workload type, as shown in *Figure 6.44*.

Figure 6.44 – Autonomous database – Workload type

We can change the workload type after database creation. But this option is available only to databases with the JSON and APEX workload type, and it can be converted only to ATP. Go to the APEX or JSON database information page and click **More Actions**. It will have the **Change Workload Type** option, as shown in the following screenshot:

Figure 6.45 – Autonomous database – Change Workload Type

Click **Change Workload Type**. We will get a window to confirm this conversion, as shown in the following screenshot.

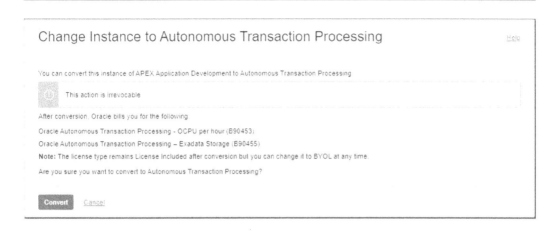

Figure 6.46 – Autonomous database – Change Workload Type

The conversion is irreversible, and the usage charges will vary after changing the workload type. When you click the **Convert** button, the automatic conversion happens in the background. After the conversion, we can see the change in the workload type of the database, as shown in the following screenshot.

Figure 6.47 – Autonomous database – workload type conversion

Enabling Operations Insights

Operations Insights analyzes the database and then provides wide-ranging insights about resource consumption, the capacity of the database, and hosts. We can get more details regarding database management from Operations Insights. This cloud-native feature is available to autonomous databases (ATP and ADB) and on-premises Oracle databases (external databases). This feature can also work on OCI Compute instances (Linux only) where the OS Management Service Agent is deployed. By default, the Operations Insights option for autonomous databases is disabled. We can find this information on the **Autonomous Database Information** page, as shown in *Figure 6.48*.

Figure 6.48 – Autonomous database – Operations Insights

Click on the **Enable** link, and you will see a pop-up window requesting confirmation, as shown in *Figure 6.49*.

Figure 6.49 – Autonomous database – Enable Operations Insights

Now click on the **Enable** button. It will enable Operations Insights and provide a link to view it, as shown in the following screenshot:

Figure 6.50 – Autonomous database – enabled Operations Insights

Click on **View**. It will open another window containing the captured details, as shown in *Figure 6.51*.

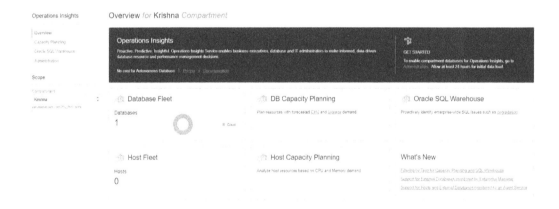

Figure 6.51 – Autonomous database – Operations Insights overview

The **Overview** page has links for **DB Capacity Planning** and **Oracle SQL Warehouse**.

The **DB Capacity Planning** page has details about CPU, memory, and storage usage, as shown in *Figure 6.52*. Currently, memory is applicable only to external databases.

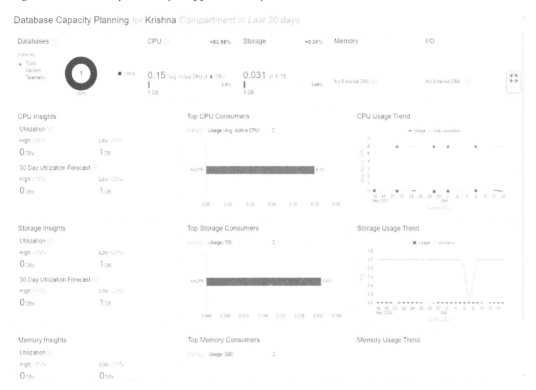

Figure 6.52 – Autonomous database – Database Capacity Planning

The analysis results are projected based on the compartment and reflect the last 30 days of data. This analysis helps us to understand resource usage trends and allows us to proactively plan and avoid unexpected capacity shortage issues. For example, as shown in *Figure 6.53*, it analyzes CPU usage and forecasts future usage. We can also get the aggregate CPU usage of all databases in the same compartment.

Figure 6.53 – Autonomous database – Database CPU details

As well as CPU forecasts, we can also forecast storage usage, as shown in *Figure 6.54*.

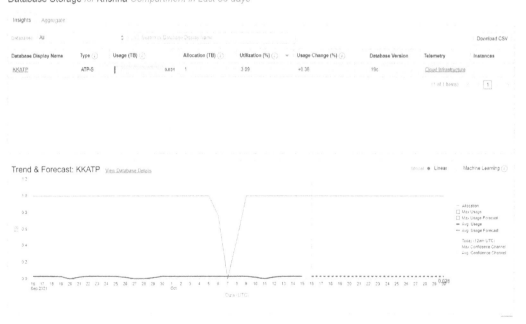

Figure 6.54 – Autonomous database – Database Storage details

The **Oracle SQL Warehouse** section provides key insights into SQL performance issues. We can identify problematic SQL statements such as degrading statements (statements with more than a 20% increase in SQL response time), SQL statements with variability (calculated based on the standard deviation of the SQL response time), inefficient statements (more than 20% inefficiency, derived from inefficient wait times), SQL statements that require improvement (statement with more than a 20% decrease in SQL response time), and SQL statements with plan changes (statements with multiple execution plans).

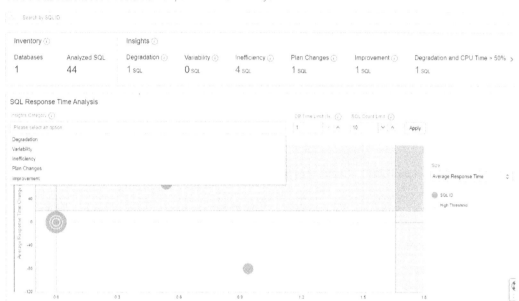

Figure 6.55 – Autonomous database – Oracle SQL Warehouse

We can also identify SQL statements with high CPU and I/O usage, as shown in *Figure 6.56*.

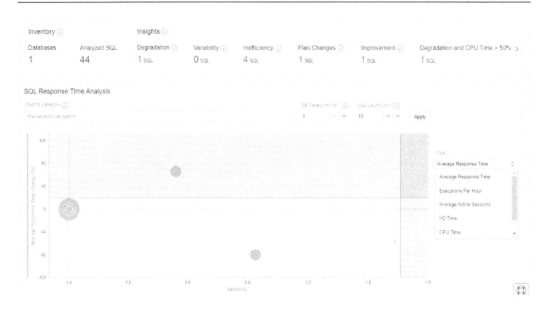

Figure 6.56 – Autonomous database – SQL Response Time Analysis

The **Administration** option on the **Operations Insights** page has links to add more databases as shown in *Figure 6.57*.

Figure 6.57 – Autonomous database – Operations Insights

It also has links to Add Hosts to Operations Insights, as shown in *Figure 6.58*.

Figure 6.58 – Autonomous database – Add Hosts

To add compute hosts, it should be with a Linux operating system image, and an Oracle cloud agent should have been installed in it. If the existing compute doesn't enable with Oracle cloud agent, then it can be installed using the Linux Yum repository:

```
yum install -y oracle-cloud-agent
```

Databases monitored through **Enterprise Manager** also can be added to Operations Insights. By adding it like this, we can perform resource analysis on databases managed by Enterprise Manager. The database could even be an external database. Basically, the data collected by Enterprise Manager targets and the **Oracle Management Repository** (**OMR**) can be transferred to OCI object storage and from there, Operations Insights can pick the data to analyze. These operations can be made automatic, whereby Operations Insights will pick the latest data.

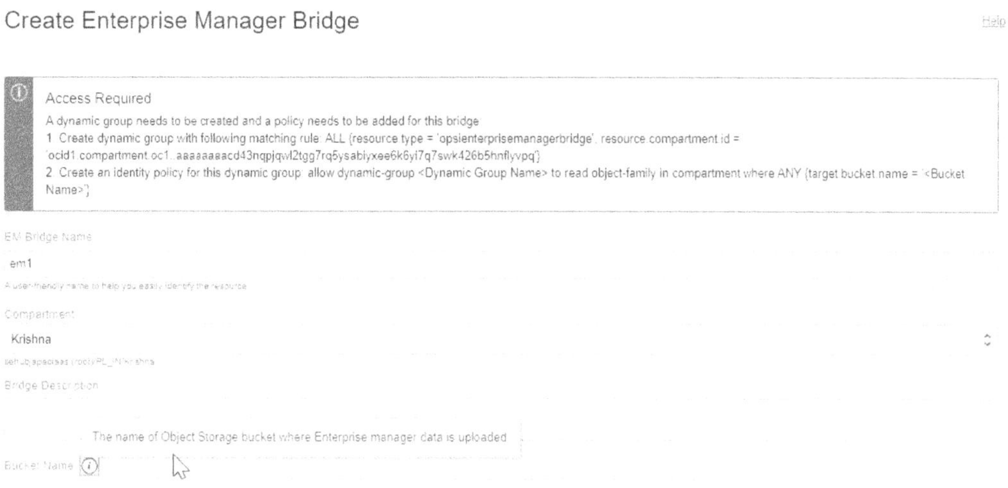

Figure 6.59 – Autonomous database – Enterprise Manager Bridge

Terminate

The terminate option drops the autonomous database. Before performing a drop operation, additional confirmation will be required. We need to provide the name of our autonomous database as confirmation to proceed with the termination.

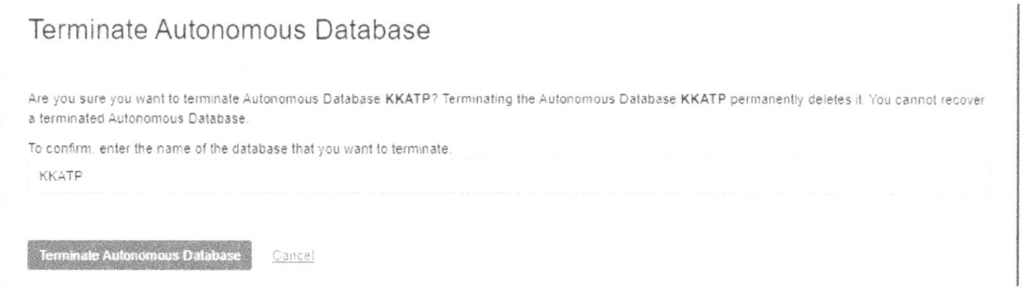

Figure 6.60 – Autonomous database termination

Registering autonomous databases with Oracle Data Safe

Oracle Data Safe provides a unified set of critical data security services for Oracle databases, whether they are running in the Oracle Cloud, on-premises, or in third-party clouds. In general, when our database is on the internet there is always the possibility of getting hacked, and therefore there is the possibility of a data breach. Data Safe overcomes all unexpected threats. Data Safe identifies database configuration drifts through overall security valuations, which helps to identify gaps and take corrective action. It flags dangerous users or behavior seen in the database and checks whether they are controlled. It audits user activity, tracks unsafe actions, raises alerts, rationalizes compliance checks, and also masks sensitive data.

In general, we need to perform a few prerequisite tasks in the database before registering it with Data Safe. But for autonomous databases, the Data Safe service account (DS$ADMIN) is pre-created. The account will be initially locked with a password expired status. When we register the autonomous database with Data Safe, OCI unlocks this account and resets its password.

To use Data Safe with an autonomous database, we first need to register the autonomous database. On the **Autonomous Database Information** page, we can find a section related to Data Safe.

Data Safe ⓘ

Status: Not Registered <u>Register</u>

Figure 6.61 – Autonomous database – Registering with Data Safe

Click **Register**, and it will ask for confirmation.

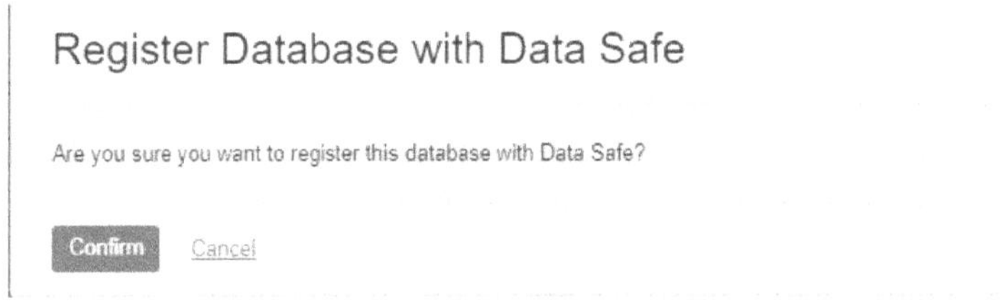

Register Database with Data Safe

Are you sure you want to register this database with Data Safe?

Confirm <u>Cancel</u>

Figure 6.61 – Autonomous database – Data Safe registration confirmation

Click **Confirm** to proceed. It will take a few minutes to complete the registration. After successful registration, we should see the **View** link, as shown in the following screenshot.

Data Safe ⓘ

Status: Registered ⎘ <u>View</u> <u>Deregister</u>

Figure 6.63 – Autonomous database – Data Safe Registered status

Click **View**, which will take us to the Data Safe page, as shown in the following screenshot.

Figure 6.64 – Autonomous database – Data Safe overview

Data Safe analyzes the database and assesses security and users. Let's discuss the options one by one.

The first option to discuss is **Targets**. Click the target at the top. It shows the screen shown in the following screenshot.

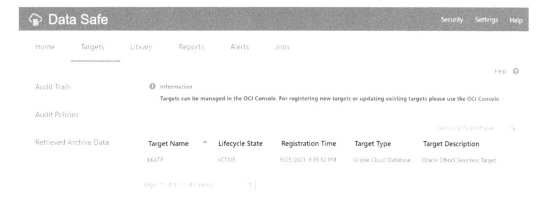

Figure 6.65 – Autonomous database – Data Safe Targets

We can find the registered autonomous databases as target names there. The auditing feature audits the user activity on the registered database so that we can monitor database usage and be alerted about unusual database activities. To view audits and alerts relating to the database, we need to provide audit and alert policies. To do that, click **Home** and then **Activity Auditing** as shown in the following screenshot.

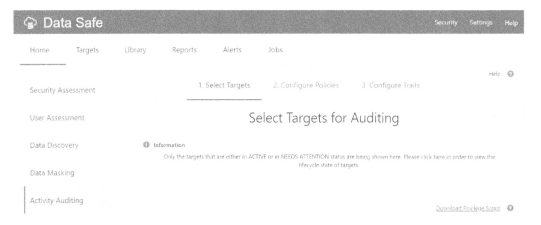

Figure 6.66 – Autonomous database – Data Safe Activity Auditing

We can see the list of target databases in the following screenshot. Choose the autonomous database and click on the **Continue** button.

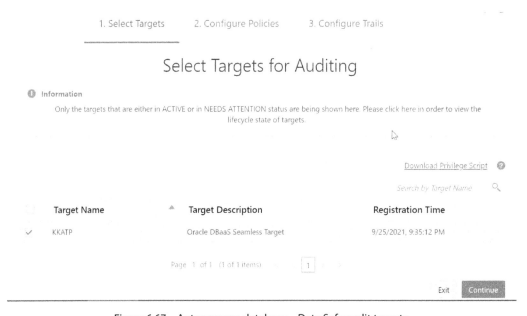

Figure 6.67 – Autonomous database – Data Safe audit targets

The next page is for retrieving audit policies. Check the checkbox for the autonomous database and click the **Retrieve** button to retrieve the audit policies for the database. We need to wait until a green check mark is displayed in the **Retrieval Status** column, as shown in the screenshot.

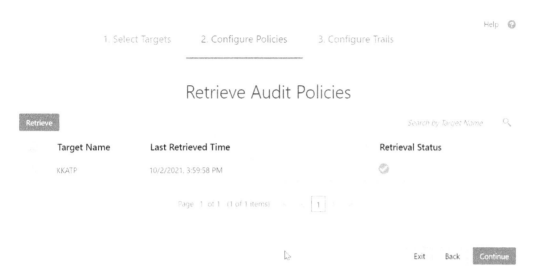

Figure 6.68 – Autonomous database – retrieving audit policies

Click the **Continue** button. The policies will be configured, and it will display the default policies configured for the database.

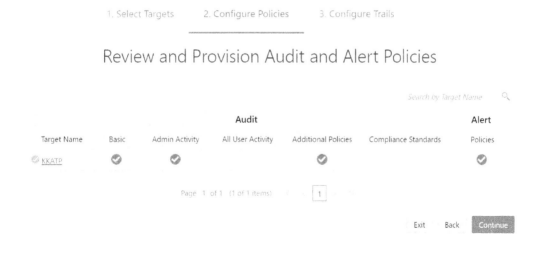

Figure 6.69 – Autonomous database – Data Safe audit and alert policies

In addition to the default policies, additional policies can be configured. Click the target database name. It will list the available **Audit Policies** and **Alert Policies**, as shown in the following screenshot.

Figure 6.70 – Autonomous database – configuring an audit policy

Oracle has many pre-defined policies, as shown in *Figure 6.71*. We can choose any policies that we require.

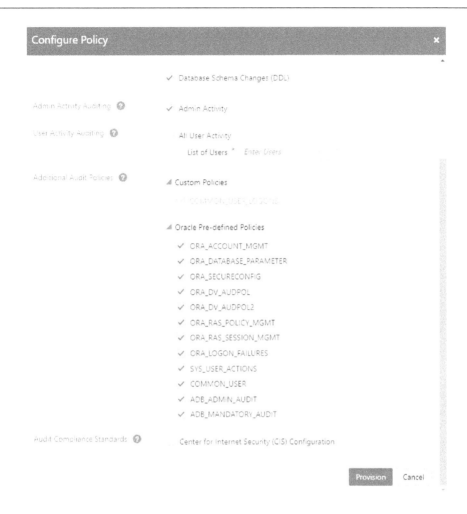

Figure 6.71 – Autonomous database – configuring audit policies

The **Critical Database Activity** policy allows us to audit critical database activities, for example, when a user, role, or profile is created, modified, or dropped.

The **Login Events** policy tracks all users' (Oracle-maintained users and custom users) login and logoff activities.

The **Database Schema Changes (DDL)** policy tracks all **Data Definition Language** (DDL) commands issued by any database user, for example, when a table, view, or trigger is created, modified, or dropped.

The **Admin Activity** policy lets you audit all activities by privileged administrators.

We could also compare the settings with **Center for Internet Security (CIS)** standards. CIS is a world-recognized organization that provides consensus-based best practices for helping organizations assess and improve their cyber security posture. The **CIS Recommendations** policy is a pre-defined unified audit policy in Oracle Database designed to perform audits that the CIS recommends. On the next tab, we can configure alert policies, as shown in the following screenshot.

Target Name :   KKATP

Audit Policies   Alert Policies

	Alert Name	Severity Level	Description
☑	Profile Changes	Critical	Changes in user profile
☑	Failed Logins by Admin User	Critical	Failed admin user login attempts
☑	Audit Policy Changes	High	Changes in audit policy
☑	Database Parameter Changes	High	Database parameter changes
☑	Database Schema Changes	Medium	Changes in database schema
☑	User Entitlement Changes	Medium	User entitlement changes
☑	User Creation/Modification	Medium	Creation or modification of users

Provision   Cancel

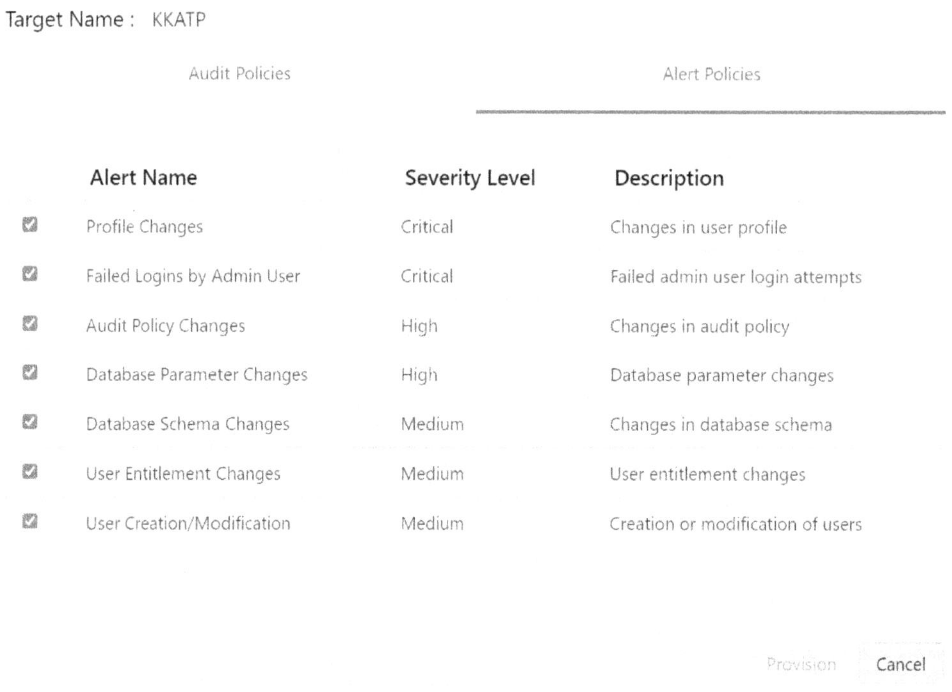

Figure 6.72 – Autonomous database – Data Safe alert policies

Alert policies define which event alerts should be invoked and their severity levels. All alert policies are rule-based, and an alert gets triggered when the event occurs that will be displayed on the **Alerts** page. Click **Provision** to enable the desired alert policies.

The next section is **Configure Trails**. This is where we start the audit collection. If audits have not been collected so far, then the state is **Not Started**, as shown in the following screenshot.

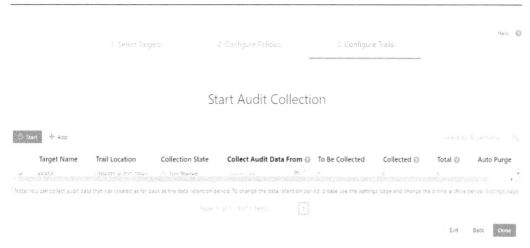

Figure 6.73 – Autonomous database – Data Safe audit collection

We need to first define from which date the audit data should be collected. Choose a date using the calendar icon available in the **Collect Audit Data From** column. Once the date and time have been chosen, it will start calculating how much audit data has to be collected, as shown in *Figure 6.74*.

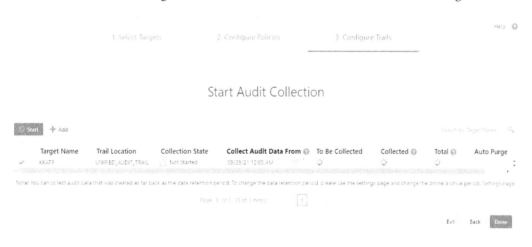

Figure 6.74 – Autonomous database – starting audit collection

Once the calculation is done, the **To Be Collected** column is updated with a value.

Figure 6.75 – Autonomous database – collecting audits

When the **To Be Collected** column contains a value, click the **Start** button to start collecting audit data.

We will be asked for confirmation before starting to collect audit data, as shown in the following screenshot.

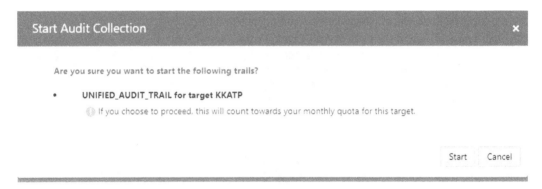

Figure 6.76 – Autonomous database – starting audit trails

A common challenge with auditing is purging the data at regular intervals. Data Safe provides an **Auto Purge** trail option. This option is available on the right-hand side, as shown in the following screenshot.

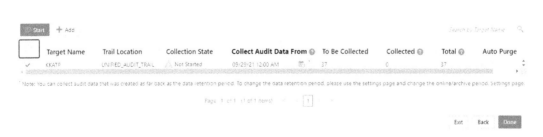

Figure 6.77 – Autonomous database – Auto Purge trail

Click the **Done** button in the **Auto Purge** trail column. It will be asked for confirmation.

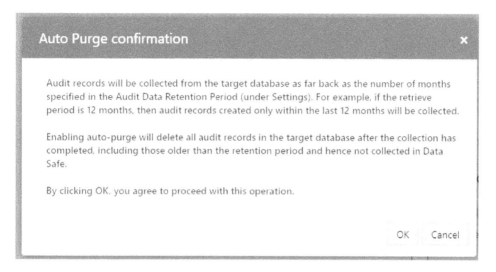

Figure 6.78 – Autonomous Database – Auto Purge confirmation

Click **OK** to initiate auto-purge. Once the audit data is collected, the **Collection State** column changes to **IDLE**, as shown in the following screenshot.

Figure 6.79 – Autonomous database – Collection State

The collected audit data can be viewed through reports. Various reports are available, as shown in the following screenshot.

Figure 6.80 – Autonomous database – Data Safe Summary details

Many reports are generated using audit data. We shall discuss Security Assessment and User Assessment reports in upcoming sections.

Security Assessment

The **Security Assessment** report lists the security risks seen in the database. On the reports page, click **Security Assessment**. This categorizes the security risks with different levels, as shown in *Figure 6.82*. Each risk type has a unique color.

Comprehensive Assessment

Category	High Risk	Medium Risk	Low Risk	Advisory	Evaluate	Pass	Total Findings
User Accounts	0	0	3	0	1	7	11
Privileges and Roles	0	0	0	0	15	5	20
Authorization Control	0	0	0	2	0	0	2
Fine-Grained Access Control	0	0	0	4	1	0	5
Auditing	0	0	0	2	8	1	11
Encryption	0	1	0	0	0	0	1
Database Configuration	0	0	0	0	3	5	8
Total	0	1	3	8	28	18	58

Figure 6.81 – Autonomous database – Data Safe – Security Assessment

We can generate the report with all the findings using the **Generate Report** button. We can also download the report for reference. The findings are categorized into multiple sections, such as **User Accounts** and **Privileges** and **Roles**. Each section details the security finding and its risk level. For example, the database has inactive users in the database, which is a low security risk, as shown in *Figure 6.82*.

Figure 6.82 – Autonomous database – Data Safe – Security Assessment

User Assessment

Let's discuss the **User Assessment** report. It validates user privileges and reports the risks associated with user settings.

Figure 6.83 – Autonomous database – Data Safe – User Assessment

As shown in *Figure 6.83*, the risks are categorized as **Low**, **Medium**, **High**, and **Critical**. Alerts are grouped by user. We can also find a table that contains the **User Name** information and all alerts related to that user. If we click on **User Name**, we can find all alerts related to that user, as shown in *Figure 6.85*.

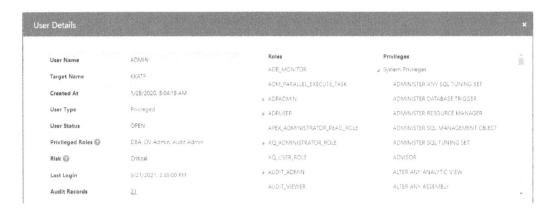

Figure 6.84 – Autonomous database – Data Safe – User Assessment details

The preceding screenshot shows alerts relating to the ADMIN user. There are 21 audit records available for this user to review. We can view details about the audit records by clicking on the audit record number.

The following figure displays a few audit records. This particular record details a login failure with the ADMIN user.

Figure 6.85 – Autonomous database – Data Safe – login failure details

We have discussed SecurityAssessment and **User Assessment** reports in this section, but there are many more reports to view. Can we view all reports in one place? Yes, we the have option to view all the activities.

In the **Reports** section, click **Activity Auditing** and **All Activity**. It will display details about all the activities.

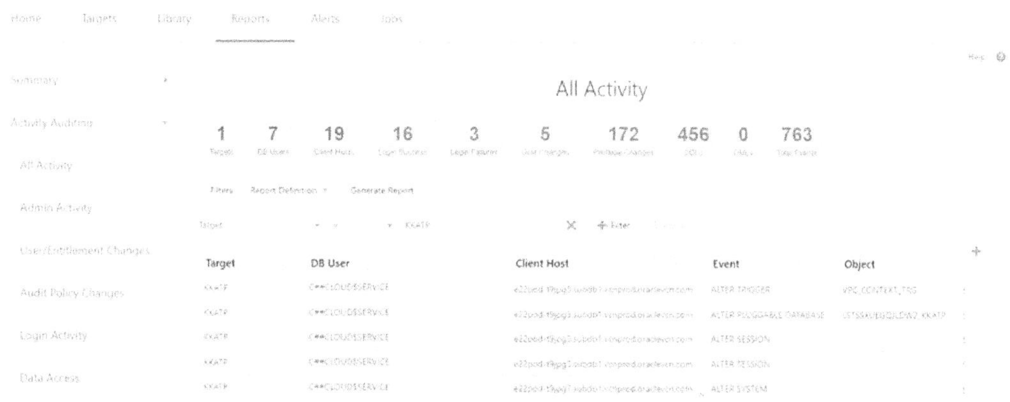

Figure 6.86 – Autonomous database – Data Safe – Reports

As shown in *Figure 6.86*, we can see details of all the activities. This page totals up the number of occurrences of each activity and projects them on that page. If we click on the number, we can see details about that activity. For example, if we click **Login Failures**, we will see the User Assessment report that we discussed in the previous session.

Like audits we can also view all alerts captured for this database. Click the **Alerts** tab next to **Reports**.

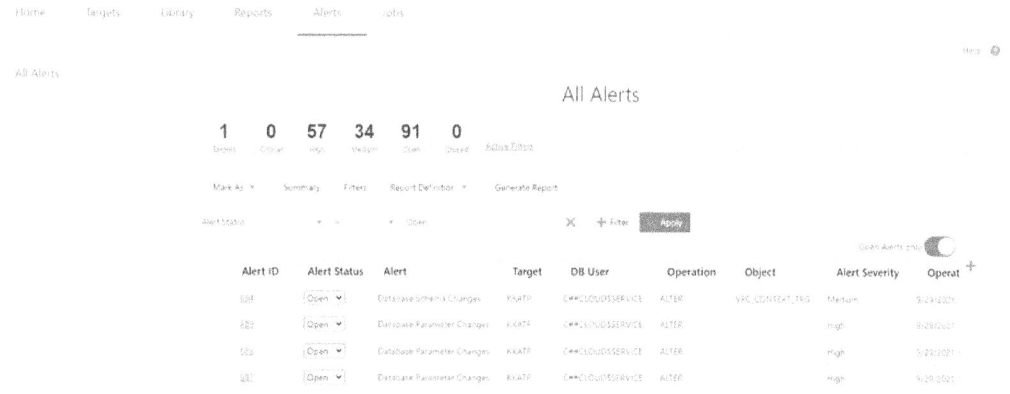

Figure 6.87 – Autonomous database – Data Safe – All Alerts

As shown in the figure, the **Alerts** page displays all alerts along with details. By clicking any alert ID in the table, we can get more details about the alert.

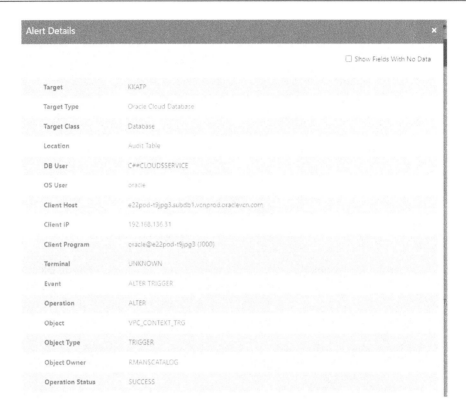

Figure 6.88 – Autonomous database – Data Safe – Alert Details

The preceding screenshot shows the details of an alert ID. It contains the client IP, the kind of event, and its operation status.

This chapter has provided insights into how to effectively manage autonomous databases. In the next chapter, we will discuss the security aspects of autonomous databases.

Summary

This chapter provided insights into all the management options for autonomous databases. We have explored how to start/stop a database, scale it up/down, clone it, and manage it. Also, we have discussed the performance metrics collected by the autonomous database and how they can be interpreted. The topics that we have discussed in this chapter will be required for day-to-day administrative activities. We now understand which tasks are automated and where manual effort is required.

In the next chapter, we will discuss a very important topic: autonomous database security. Most security attributes will be taken care of automatically, but users may need to contribute in a few places. We will discuss this in detail in the next chapter.

Part 3 – Security and Compliance with Autonomous Database

You will learn about the security aspects of Autonomous Database, such as customer-owned keys for encryption, Data Safe, and managing keys in OCI Vault.

This part comprises the following main chapter:

- *Chapter 7, Security Features in Autonomous Database*

7

Security Features in
Autonomous Database

Security is always one of the high-priority topics to discuss in terms of the cloud environment. With respect to Autonomous Database, it is very interesting to discuss the security aspects to understand how security is managed automatically. Since an autonomous database is a self-driven database, it's also exciting to see how it automatically ensures the database is safe from all outside attacks. In this chapter, we are going to discuss Autonomous Database infrastructure security, and then security related to data management such as Data Safe and Activity Auditing. Let's discuss all the security topics in detail one by one.

The agenda of this chapter includes the following topics:

- Operating system access restriction
- Security Assessment
- Data Discovery
- Data Masking and Activity Auditing
- Customer-managed keys with Autonomous Database

Let's begin!

Operating system access restriction

Usually, in a database environment, the **database administrator** (**DBA**) accesses the server to install and manage the database. They will create the necessary data files and then load the data into the database. DBAs and system administrators also access the servers to perform day-to-day maintenance tasks, such as database backup generation, archive logs generation, data files growth, and rectifying physical corruption. They also access the operating system structure to perform the scaling of the CPU, memory, and storage. Here, in the Autonomous Database, the provisioning and management are automated. All database management tasks are taken care of automatically. Hence, Autonomous

Database only provides database connectivity. The underlying operating system is not accessible from the outside. We can't connect the database server via SSH. The Autonomous Database server is secured and managed by Oracle.

Not allowing access to the operating system makes this environment more secure. Unwanted operations to the operating system are avoided. In general, when performing database maintenance tasks, there are chances of human mistakes being made, such as mistakenly deleting files (data files or archived logs) or sometimes forgetting to purge unneeded files (obsolete backups or archived logs) after their use. These challenges are completely avoided in an autonomous environment. All database admin-related activities are taken care of by the autonomous database. Zero human interaction reduces the opportunity for human error.

Certain Oracle utilities require server access, such as Data Pump, which is a server-side utility. Data Pump generates dump files and log files in the database server. You may ask "If I don't have server access, how will I generate a dump file using this utility?" For the autonomous database, the dump file can go directly into the Object Storage bucket, instead of into the server. Logs will be stored in the database server anyhow, but they can be copied to the Object Storage bucket using the DBMS_CLOUD package.

DBMS_CLOUD is a PL/SQL package that comes along with Oracle Database. It provides complete support for working with data in OCI Object Storage.

In the following example, export.log is generated by Data Pump and stored in the autonomous database server. We can move it to Object Storage using the DBMS_CLOUD package:

```
BEGIN
  DBMS_CLOUD.PUT_OBJECT(
    credential_name => 'DEF_CRED_NAME',
    object_uri => 'https://objectstorage.<region>.oraclecloud.
com/n/namespace-string/b/bucketname/o/import.log',
    directory_name  => 'DATA_PUMP_DIR',
    file_name => 'export.log');
END;
```

export.log will be in DATA_PUMP_DIR by default. Since we don't have access to the autonomous database server, we can execute the preceding procedure to move the log file to Object Storage.

Dedicated ADMIN users for the database

Autonomous Database restricts SYS and SYSTEM schema access. These are high-privileges users in the database with the right to modify the database structure. Since the autonomous database is a managed database, these users are restricted from login. The SYSDBA privilege is also restricted, which means we can't grant the SYSDBA privilege to other users. If we try to grant the SYSDBA privilege, we will get an error as shown here:

```
SQL> Grant sysdba to admin;
Error starting at line : 8 in command -
Grant sysdba to admin
Error report -
ORA-01031: insufficient privileges
01031. 00000 -  "insufficient privileges"
```

Instead of SYS or SYSTEM users, Autonomous Database provides an ADMIN user. The ADMIN user has a PDB_DBA role. The PDB_DBA role is applicable to multitenant environments. By default, Autonomous Database comes with a multitenant model. A non-multitenant model has been deprecated since Autonomous Database version 12.2. ADMIN can create additional schemas and add or revoke privileges. The ADMIN user password can be reset at any time through the OCI portal, as shown in the screenshot:

Figure 7.1 – Changing the administrator password

By clicking the **Administrator Password** option, a new window for specifying the password details will open, as shown in *Figure 7.2*:

Administrator password   Help

Change your administrator's password.

Username *Read-Only*

ADMIN

Password

Confirm password

Change   Cancel

Figure 7.2 – Specifying a new password for the administrator user

The password should follow a few criteria:

- Contains from 12 to 30 characters
- Contains at least one lowercase letter
- Contains at least one uppercase letter
- Contains at least one number
- Does not contain the double quotation mark (")
- Does not contain the admin string, regardless of its case
- Is not one of the last four passwords used for the database
- Is not a password you previously set within the last 24 hours

Once the password is provided, click on the **Change** button, as shown in *Figure 7.2*. It will take a few seconds to modify the password.

Private endpoint

Autonomous Database provides the option to define the mode of SQLPLUS connectivity. It allows three SQLPLUS connectivity options, such as connectivity from the internet, connectivity from specific IPs or VCNs, and private endpoints.

On the autonomous database details page, click **More Actions**. On there, you'll see the **Update Network Access** option, as seen in the following screenshot. Click on that:

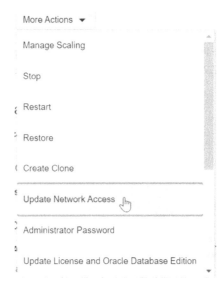

Figure 7.3 – Updating network access

A new window will open requesting you to choose the desired network access. **Private endpoint access only** is one of the options available to choose, as shown in *Figure 7.4*:

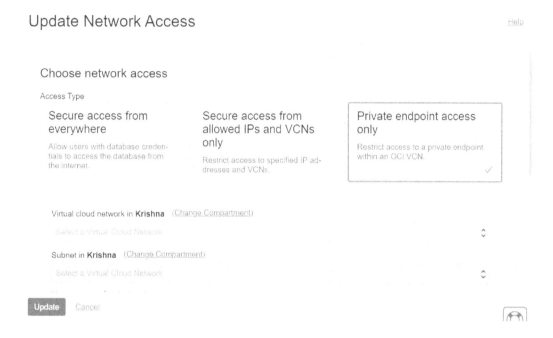

Figure 7.4 – Private endpoint access only

A private endpoint connection makes Autonomous Database connectivity secure. It only allows chosen OCI VCN and subnet CIDR ranges to connect with the autonomous database. It also needs a **Network Security Group (NSG)** to restrict the protocol and port to make the connection. Let's learn more about it in the next subsection.

NSG

In the private endpoint, we can restrict the database access to a particular VCN or subnet. With an NSG, we can filter traffic to a specific VM in that VCN or a specific IP or CIDR range.

An NSG is similar to a security list that consists of a set of ingress and egress rules. In the security list, the rules are applied at the subnet level, whereas NSG rules are applied only to a set of VNICs for a particular VCN. Consider you have multiple VMs in a subnet and you want to block a few ports on certain VMs – this can be achieved via an NSG.

In OCI, an NSG can be created from the VCN page. On the **VCN details** page, under the **Resources** section, you can find **Network Security Groups**, as shown in *Figure 7.5*:

Figure 7.5 – NSGs

Click on the **Network Security Groups** link. This brings up the page where we can find the button to **Create Network Security Group**, as shown in *Figure 7.6*:

Figure 7.6 – Creating an NSG

Now, click on the **Create Network Security Group** button. It will invoke the window shown in *Figure 7.7*:

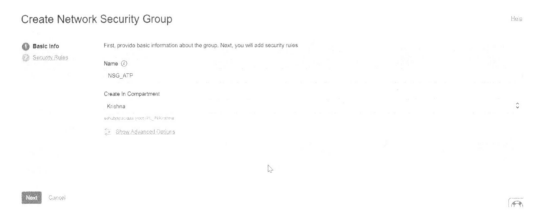

Figure 7.7 – Basic info to create an NSG

Enter a given name and the **Create In Compartment** details, which dictate where the NSG needs to be created. On the next page, we can find the security rule fields for the NSG, as shown in *Figure 7.8*:

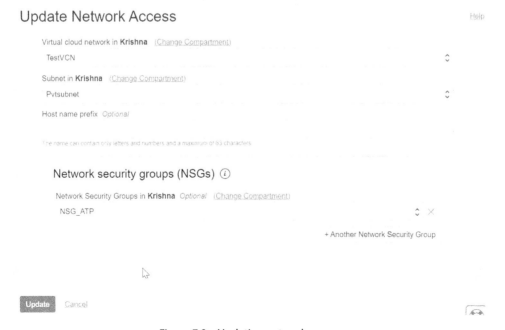

Figure 7.8 – Security rules for the NSG

Now, provide the details of the source connection and create the NSG.

This NSG can be added to the autonomous database private endpoint. On the **Private endpoint** page, choose the VCN and subnet, and right under it is a field in which to pick the NSG. Here, we can choose the NSG that we have created in that VCN as shown in *Figure 7.9*:

Figure 7.9 – Updating network access

Once the details are provided, click on the **Update** button to complete the private endpoint configuration.

Data Safe

Data Safe is a tool provided by OCI for analyzing the database to ensure its security. With data breaches growing every day, along with an evolving set of data protection and privacy regulations, protecting business sensitivity and regulated data became mission-critical. The Data Safe tool analyzes databases, whether on-premises or cloud databases (Database Cloud Service or Autonomous Database or databases on a bare metal server, compute, or Cloud@Customer), in terms of their configuration, operation, and implementation and then recommends changes and controls to mitigate any risks.

Data Safe identifies configuration settings that may increase risk exposure and also identifies sensitive user accounts, their entitlements, and security policies. An autonomous database also can be analyzed by Data Safe.

First, the database needs to be registered with Data Safe. To register the databases, they should be accessible to OCI. This applies whether it is on-premises or on the cloud. We will discuss the prerequisites with corresponding screenshots in detail:

1. In the OCI portal, the **Data Safe** service is available under the **Oracle Database** section, as shown in *Figure 7.10*:

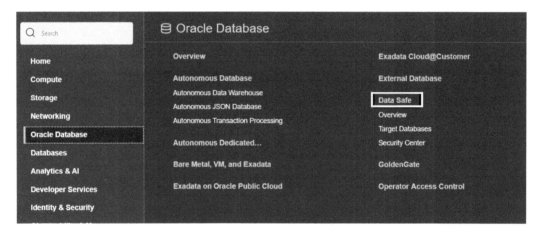

Figure 7.10 – Data Safe in the OCI portal

2. By clicking on the **Data Safe** link, we will come to the **Overview** page:

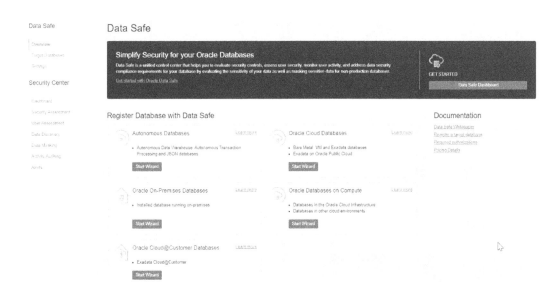

Figure 7.11 – Overview of Data Safe

3. To analyze a database, the first step is to register it with Data Safe. This registration can be done using the **Target Databases** link or the wizard.

Using the wizard, it's a five-step process:

Figure 7.12 – Selecting a database

The actions performed in this step collect the details of the autonomous database. Based on the network setup, the connectivity and security rule options will then be enabled. If the database is open for public access, then there is no need to specify the connectivity or security rule options. After reviewing the details, click on the **Register** button.

4. Another way we can register the database is by clicking the **Target Databases** link available on the **Overview** page. Click on the **Target Databases** link and then click on the **Register Database** button on the **Target Databases** page:

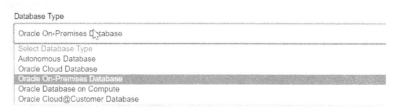

Figure 7.13 – Choosing the database type

It accepts **Autonomous Database**, **Oracle Cloud Database**, **Oracle On-Premises Database**, **Oracle Database on Compute**, and **Oracle Cloud@Customer Database**. Each option has different arguments for making a connection to the database. For example, an on-premises database requires a connection descriptor that has details about the database server and port. The database should be reachable by OCI.

5. Choose **Autonomous Database** from the dropdown. Make sure the database has been started already. In the **Data Safe Target Information** section, we can find all the active autonomous databases. Choose the desired database and then provide a name under **Data Safe Target Display Name**. A description is an optional addition and a given **Compartment** for the target database can be chosen, as shown in *Figure 7.14*:

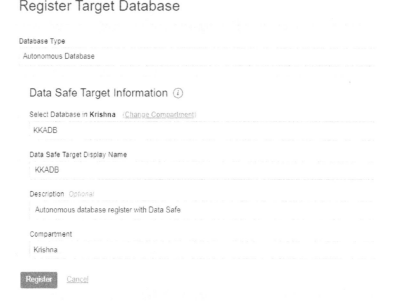

Figure 7.14 – Registering a target database

6. After providing all the details, click on the **Register** button. The database will be registered with Data Safe. The **Target Database Details** page will be displayed after a successful registration, as shown in *Figure 7.15*:

Figure 7.15 – Target database after registration

Essentially, Data Safe connects with the database through a schema called DS$ADMIN in Autonomous Database. This user is created by default in Autonomous Database and this user is going to be permanent. Hence, we see the **Update Database User** button is disabled in the preceding screenshot. For other database flavors, this button will be enabled.

If we are registering a non-autonomous database, then before registering the target database, we need to create a DATASAFE$ADMIN schema there. This schema performs the necessary collection from the target database and for that, it requires a few privileges. The required privileges are scripted into an SQL file and are available to download on the registration page itself. We can download and execute it in the target database.

Once the target database is registered, we can move to the Data Safe Security Center where we can view the results after analysis. For an autonomous database, we don't need to download and execute the SQL file – here, everything is automated.

The Security Center shows the collected details in different sections, as shown in *Figure 7.16*:

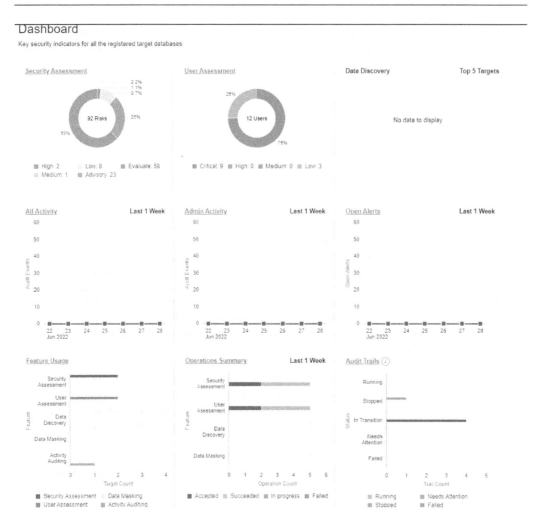

Figure 7.16 – Data Safe dashboard

The overview shows details of all sections with respect to all registered target databases. If we want the details of a particular database, then the **Filters** option can be used. The **Filters** option is available on the left-hand side after **Compartment**, as shown in *Figure 7.17*:

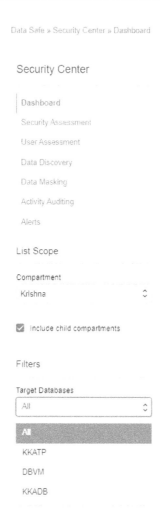

Figure 7.17 – Target databases list in Data Safe

Let's discuss each section of the Security Center in detail.

Security Assessment

In **Security Assessment**, Data Safe evaluates the security posture of the databases and provides recommendations to overcome any issues. It enables us to identify security vulnerabilities and verify that encryption, auditing, and access controls have been implemented. It categorizes the findings into different sections, as shown in *Figure 7.18*:

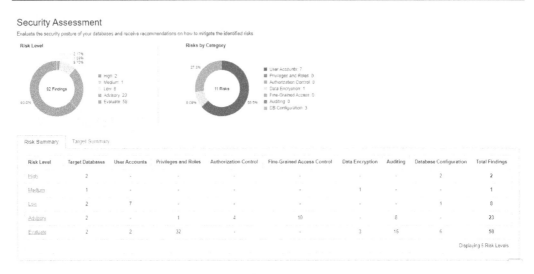

Figure 7.18 – Security Assessment

As you can see in *Figure 7.18*, the risks are categorized into different sections:

- **User Accounts**
- **Privileges and Roles**
- **Authorization Control**
- **Fine-Grained Access Control**
- **Auditing**
- **Data Encryption**
- **Database Configuration**

It also categorizes the risk level into five categories based on the severity involved:

- **High**: An item that needs immediate attention
- **Medium**: It is not very urgent, but planning is required to address it in the near future
- **Low**: It's a low-severity risk and can be taken care of in the next maintenance window
- **Advisory**: It's recommended to address it in order to improve the security posture
- **Evaluate**: It requires manual analysis

All risk levels have a link for showing the details. By clicking on **Risk Level**, we can find the risk details and the associated registered databases:

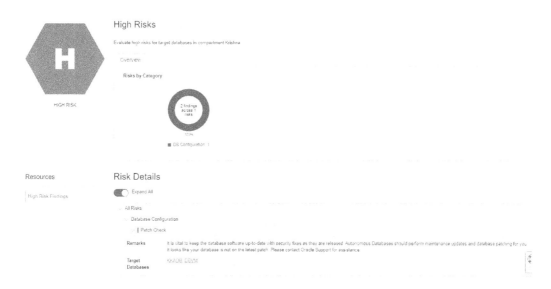

Figure 7.19 – High risks in Security Assessment

Figure 7.19 shows the details of a high-level risk related to the patch level. The patch of the database software is not up to date. This alert has been reported for the **KKADB** and **DBVM** databases. By clicking on the database, we can find out more details on the alert:

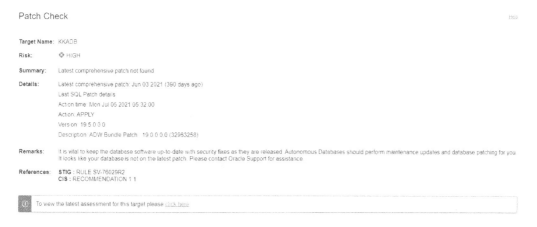

Figure 7.20 – High risk

In this example, the autonomous database was down for a long time – hence, the patch is not up to date. The patching will be taken care of automatically in the next maintenance period.

We can also check all the alerts for a particular registered database by clicking on the **Target Summary** option, as shown in the following figure:

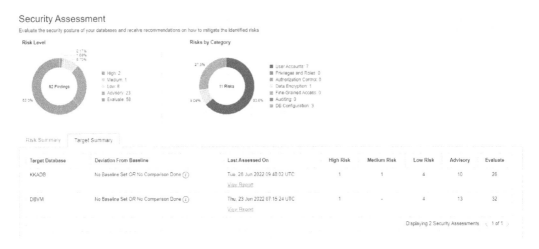

Figure 7.21 – Target Summary in Security Assessment

By clicking on **View Report**, as shown in *Figure 7.21*, we are able to see all the **Security Assessment** information for a registered database.

User Assessment

User Assessment collects reports by analyzing the database user accounts and helps to identify potential risks present in the database user accounts. It assesses the user account along with its metadata, such as roles, privileges, and profiles, and calculates a risk score for each user:

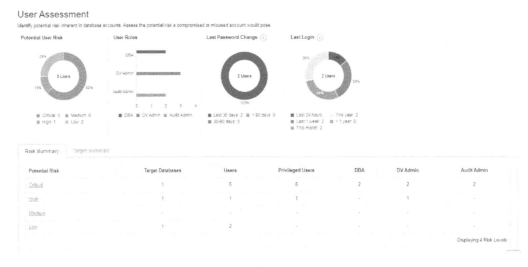

Figure 7.22 – User Assessment

The risk levels are categorized as **Critical**, **High**, **Medium**, and **Low**, as shown in *Figure 7.22*:

- **Critical**: It lists the risks at the database level that have an impact on database availability and integrity, such as having direct access to read, modify, copy, or access data.

- **High**: It lists the risks at the feature level. It talks about the ability to read, modify, or copy data indirectly through the use of the privileges granted.

- **Medium**: It lists risks that don't have serious effort but can damage user management, such as setting lists during user sessions.

- **Low**: The risks with the least impact, such as those affecting a particular user.

By clicking on the risk level, we get to the **User Details** page, as shown in *Figure 7.23*:

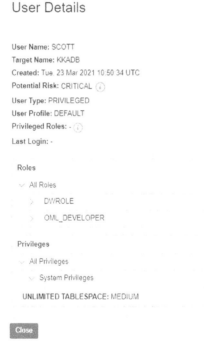

Figure 7.23 – A critical risk in User Assessment

The preceding risk talks about privileges granted to SCOTT. This user has DWROLE and OML_DEVELOPER roles and the Unlimited Tablespace privilege. DWROLE provides common privileges for a database developer and data scientist to perform real-time analytics. OML_DEVELOPER is related to machine learning. Assigning these privileges to SCOTT is considered a security threat and that has been reported as a risk.

The next section is about Data Discovery.

Data Discovery

Data Safe scans the database for sensitive data. The sensitive data will be identified based on column names, actual data, and comments. Data Safe provides over 170 predefined sensitive types that are used to search for sensitive data in the database. We can add new sensitive types to search for, but existing types can't be altered.

The top-level categories for predefined sensitive types are as follows:

- **Identification information**: Sensitive identifiers such as a US Social Security number or Indian Aadhaar number

- **Biographic information**: Sensitive types such as a complete address, date of birth, or religion

- **IT information**: Sensitive types such as a user ID, password, or IP address. Data related to user IT data

- **Financial information**: Sensitive information such as credit card information and bank account details

- **Healthcare information**: Sensitive types such as a health insurance number or blood type

- **Employment information**: Sensitive information such as an employee title, income, stock, or an employee number

- **Academic information**: Sensitive data such as a college student ID and grade or degree information

Data Safe saves the discovery results as a sensitive data model. You can find the **Discover Sensitive Data** button in the **Data Discovery** section, as shown in *Figure 7.24*:

Figure 7.24 – Data Discovery

Click on **Discover Sensitive Data** to examine the sensitive data in the database. It will invoke the **Create Sensitive Data Model** page, as shown in the following screenshot:

Create Sensitive Data Model

Figure 7.25 – Sensitive data model information

Provide the **Compartment** and **Target Database** name details on the first page, as shown in *Figure 7.25*. On the next page, we need to choose either all schemas or a particular schema to discover the sensitive data, as shown in the following screenshot:

Figure 7.26 – Selecting the schemas for the sensitive data model

On the next page, choose the sensitive types that we want to discover:

Figure 7.27 – Sensitive types

We can choose a particular type or all types, as shown in *Figure 7.27*. On the next page, we have discovery options:

Figure 7.28 – Discovery options

As shown in *Figure 7.28*, the available options are **Collect, display, and store sample data** (collect sample data from the database) or **Discover application-level (non-dictionary) referential relationships**.

Click on **Create Sensitive Data Model** as the last step.

Data Safe will start the discovery process with the provided details and a sensitive data model will be created, as shown in *Figure 7.29*:

Figure 7.29 – Sensitive data model

The data model shows that it has detected sensitive schemas, tables, and columns. We can find the details of the information on the same page, as shown in *Figure 7.30*:

Figure 7.30 – Sensitive columns

We can also carry out **Incremental Discovery** at a later point in time.

Now, if we move to the **Data Discovery** page, an overview of the collected information will be shown, as we can see in *Figure 7.31*:

Figure 7.31 – Data Discovery

We can also find the predefined sensitive types on the same page under **Related Resources,** as shown in *Figure 7.31*. Now, click on the **Sensitive Types** link:

Sensitive Types

Sensitive types define the kinds of columns to search for. They are used for discovering and classifying sensitive columns. This page lists the user-defined sensitive types and categories in the selected compartment, along with all the predefined sensitive types and categories. Learn more.

Create Sensitive Type/Category					
Name ▲	Description	Parent Sensitive Category	Sensitive Category	Oracle Predefined	
Academic Degree	Person''s educational qualification. Examples: PhD, MBA, Master of Science.	Student Basic Data	No	Yes	⋮
Academic Information	Education data that can potentially be used to identify an individual.	-	Yes	Yes	⋮
Academic Level	Academic level, class, or standard. Examples: 9th Standard, Sophomore.	Performance Data	No	Yes	⋮
Address	Physical address. Examples: Street, City, State, Country.	Biographic Information	Yes	Yes	⋮

Figure 7.32 – Sensitive types template

You can find all the predefined sensitive types on the page. In total, 178 sensitive types are displayed on 12 pages.

Data Safe also offers **Data Masking**, which we are going to discuss in the next section.

Data Masking

Data Safe masks the sensitive data that it has identified in the target databases. It provides an option to create a masking policy for the sensitive data model that we have seen in the previous section.

Click on **Masking Policies** under **Related Resources,** as shown in *Figure 7.33*:

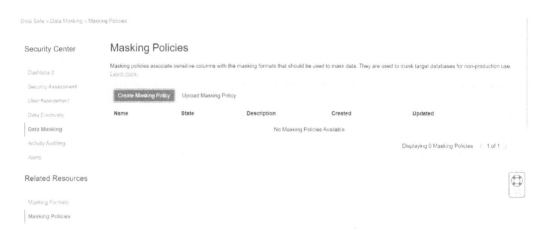

Figure 7.33 – Masking policies

Click on the **Create Masking Policy** button, as shown in *Figure 7.34*:

Masking Policies

Masking policies associate sensitive columns with the masking formats that should be used to mask data. They are used to mask target databases for non-production use. Learn more.

Create Masking Policy   Upload Masking Policy

Name	State	Description	Created	Updated
		No Masking Policies Available		

Displaying 0 Masking Policies   ⟨ 1 of 1 ⟩

Figure 7.34 – Creating a masking policy

On the **Create Masking Policy** page, provide the details of the policy name and specify whether the policy has to be created using an available sensitive data model or using an empty masking policy that is associated with the selected target database:

Figure 7.35 – Creating a masking policy using a sensitive data model

And here's how the screen looks like when you upload scripts

Figure 7.36 – Uploading masking scripts

Click on **Create Masking Policy** to create a masking policy and add columns from the sensitive data model. Once the policy has been created successfully, it will list masking columns that it has considered, as shown in *Figure 7.37*:

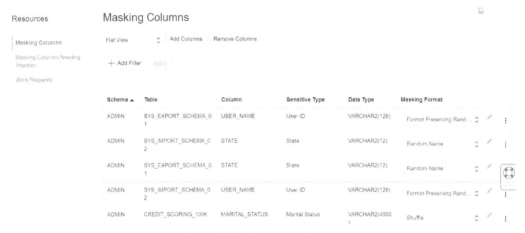

Figure 7.37 – Masking columns

Each masking column will be listed with details on the schema, table name, column name, sensitive type, data type, and masking format. Data Safe identifies a suitable masking format for each sensitive column. The masking format can be edited. Each masking format has a different kind of masking algorithm:

Edit Masking Format

Masking Column: ADMIN.SYS_IMPORT_SCHEMA_02.STATE
Sensitive Type: State
Data Type: VARCHAR2(12)
Assigned Masking Format: Random Name

Description: • Replaces values with random letters of random length. Compatible with character type columns

Examples: • 'AjHjK123#@' could become 'Sbvtud'
• 'Michael' could become 'Ramzori'
• 'Richard Williams' could become 'Madpalvik'

Condition (default: 1=1)

1 = 1

Masking Format Entry

Random Name

Library Masking Format Entry

User Defined Function

Schema Name

MASKCS_FMTLIB

Package Name *Optional*   Function Name

RANDOM_NAME

+ Another Format Entry

Figure 7.38 – Editing the masking format

Figure 7.38 shows the masking format for **Random Name**. The examples show how the masking will be carried out. It replaces the column value with different characters.

Now that the masking policy is ready, let's start masking sensitive data using the policy that we have created. Click on **Mask Sensitive Data**. It will invoke a window in which to choose the target database and masking policy, as shown in *Figure 7.39*:

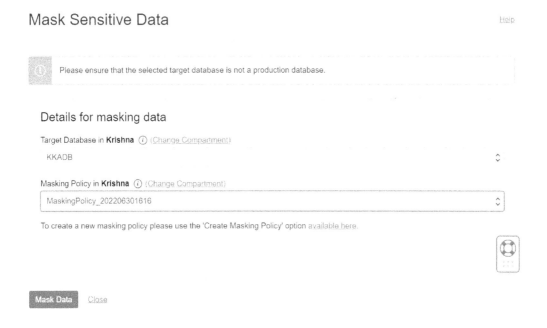

Figure 7.39 – Masking sensitive data

Click on the **Mask Data** button, which will create the masking job:

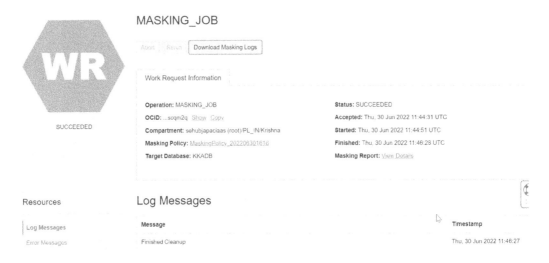

Figure 7.40 – Masking job

Download Masking Logs is available, which will generate the log as a text file. The log contains all the SQL commands and their output.

Activity Auditing

Data Safe collects audit data from target databases and helps to ensure accountability and improve regulatory compliance and monitor user activities in target databases. We can audit sensitive data, security-related events, and activities that are chosen to be monitored. As we see in *Figure 7.41*, all events are monitored and reported on the Activity Auditing page:

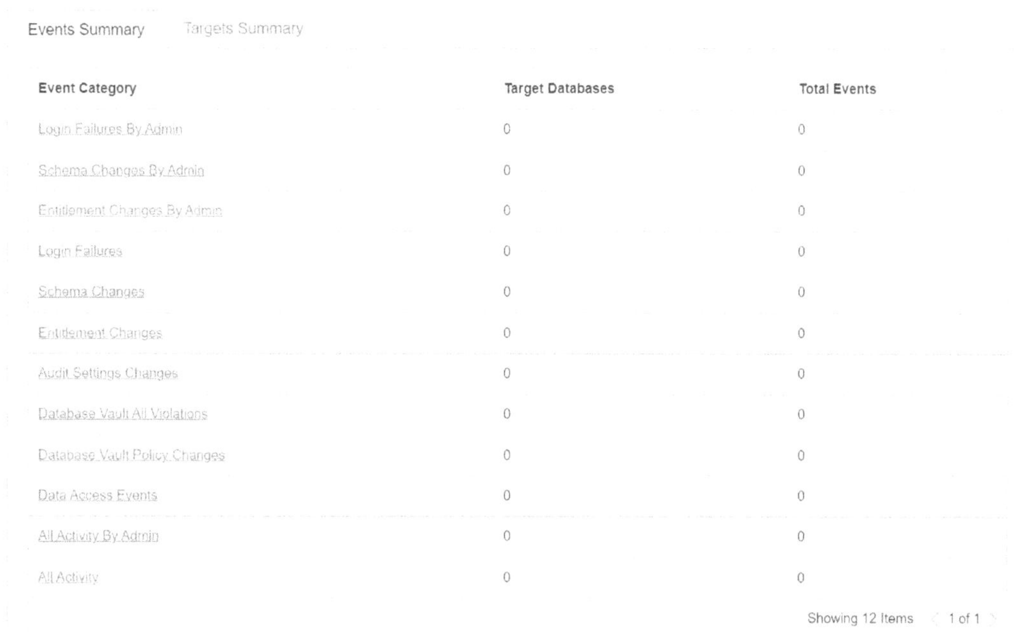

Events Summary   Targets Summary

Event Category	Target Databases	Total Events
Login Failures By Admin	0	0
Schema Changes By Admin	0	0
Entitlement Changes By Admin	0	0
Login Failures	0	0
Schema Changes	0	0
Entitlement Changes	0	0
Audit Settings Changes	0	0
Database Vault All Violations	0	0
Database Vault Policy Changes	0	0
Data Access Events	0	0
All Activity By Admin	0	0
All Activity	0	0

Showing 12 Items   1 of 1

Figure 7.41 – Auditing events

The details of each activity auditing are given here:

Report Name	Description
All Activity	All database activities will be audited.
Admin Activity	Activities performed by Administrative users.
User/Entitlement Changes	User related activities such as user creation/deletion/privilege and role changes.
Audit Policy Changes	All changes in audit policies.
Login Activity	Database login attempts.
Data Access	Database query operations.
Data Modification	Data modification activities (DMLs).
Database Schema Changes	Database schema changes (DDLs).
Data Safe Activity	Activity generated by the Oracle Data Safe service.
Database Vault Activity	Auditable activities of enabled Oracle Database Vault policies in target databases. It includes mandatory Database Vault configuration changes, realm violations, and command rule violations.

Up to 1 million audit records per month per target database are included in Oracle Data Safe at no additional cost.

We have discussed all sections of Data Safe. Let's move on to the next security feature, data encryption.

Data encryption

An autonomous database encrypts data by default at rest and in transit. It can't be turned off. There are two ways to perform data encryption. These are encryption of data at rest and encryption of data in transit. Let's discuss these.

Encryption of data at rest

All tablespaces are encrypted using **TDE** (**Transparent Data Encryption**). **TLS** (**Transport Layer Security**) 1.2 protocol has been applied. Encryption protects the processing, transmission, and storage of data. Each database and its backup have different encryption keys. Database cloning creates a new set of keys.

An autonomous database manages the keys by default and stores them in the **Public Key Cryptography Standards** (**PKCS**) 12-key store, but it also has provision to use customer-managed keys created using the OCI Vault service. The customer-managed keys can be rotated regularly to meet the best security standards.

To configure custom-managed keys for Autonomous Database, log in to the OCI portal and find the autonomous database details page. Click on the **More actions** button and then the **Manage Encryption Key** option, as shown in *Figure 7.42*:

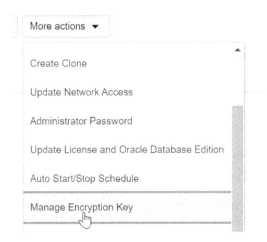

Figure 7.42 – Managing the encryption key

A new window will pop up, as shown in *Figure 7.43,* where we can specify the OCI Vault and custom encryption key to perform the encryption:

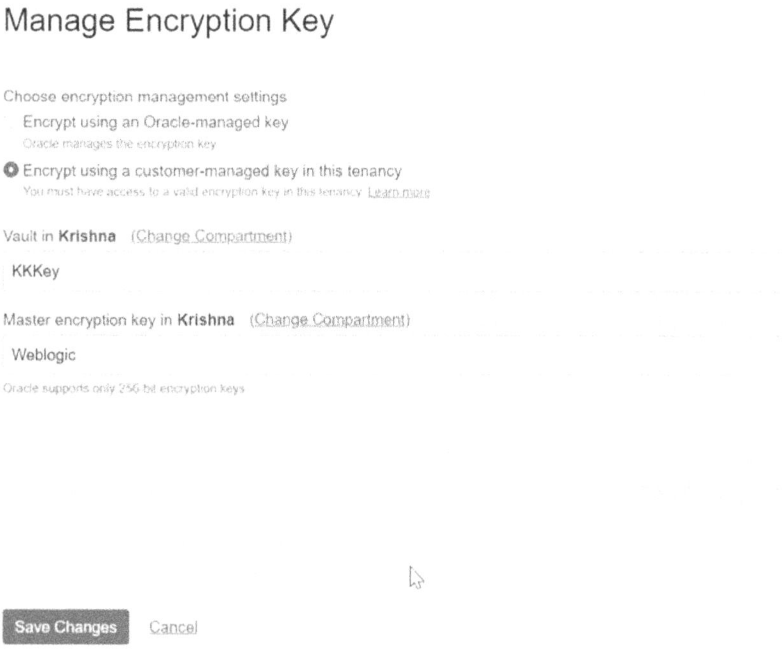

Manage Encryption Key

Choose encryption management settings

Encrypt using an Oracle-managed key
Oracle manages the encryption key

⦿ Encrypt using a customer-managed key in this tenancy
You must have access to a valid encryption key in this tenancy. Learn more

Vault in **Krishna**   (Change Compartment)

KKKey

Master encryption key in **Krishna**   (Change Compartment)

Weblogic

Oracle supports only 256-bit encryption keys

Save Changes   Cancel

Figure 7.43 – Custom encryption key

From the drop-down box, we can choose the vault first. This will populate the keys associated with that vault and list them in the master encryption key drop-down box. We can choose the key and click on the **Save Changes** button:

Encryption of data in transit

As we discussed earlier, Autonomous Database only provides the database connections and not server access. The connection could be SQL*Net, JDBC, or ODBC.

The autonomous database supports **Mutual Transport Layer Security (mTLS)** and TLS connections. By default, mTLS connections are supported. We can find the details on the autonomous database page, as shown in *Figure 7.44*:

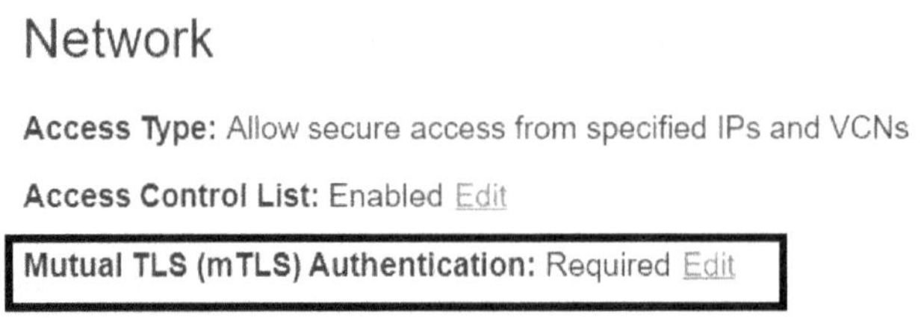

Network

Access Type: Allow secure access from specified IPs and VCNs

Access Control List: Enabled Edit

Mutual TLS (mTLS) Authentication: Required Edit

Figure 7.44 – TLS authentication

mTLS clients connect through a TCPS (secure TCP) database connection using standard TLS 1.2 with a trusted client **Certificate Authority** (**CA**) certificate. The mTLS connections require a wallet to be downloaded to make a connection with the database. In *Chapter 5*, *Backup and Restore with Autonomous Database in OCI*, we discussed the wallet and how to download and make connections using it. Users who have the wallet can make connections to the database along with user authentication. The data encryption happens via the wallet. With mTLS connections, the wallet is located at both the client and the database server. The certificate authentication uses the key stored in the wallet to do the encryption. The key needs to match the client and the server to make a connection.

TLS clients connect through a TCPS database connection using standard TLS 1.2. A client uses its list of trusted certificates to validate the server's CA root certificate. TLS connections are used by a JDBC Thin driver client, which doesn't need to download the wallet.

Database backup encryption

Database backups are also encrypted. The autonomous database takes a backup of the database automatically. It also has a provision to make database backups manually. We discussed the database in detail in *Chapter 5*. From a security point of view, we can highlight that all database backups are encrypted by default in an autonomous database. The encryption keys are rotated by default.

We can verify the backup encryption using the following command:

```
SQL> select count(*) from v$backup_set_details where
encrypted='NO' order by completion_time desc;
Count(*)
-------------
0
```

While decrypting the backup, the wallet is mandatory. The wallet also will be backed up during the regular backup process.

Access control lists

An **Access Control List** (**ACL**) configures an autonomous database that only allows traffic from allow-listed source applications, systems, and individual IP addresses. By default, all traffic is blocked by the ACL. It protects data from random user queries, unknown or uncertified applications, or user access:

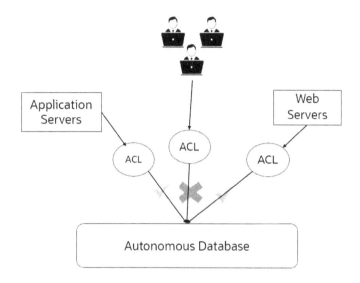

Figure 7.45 – ACL

We can find whether the ACL is enabled at the database level on the autonomous database details page, as shown in *Figure 7.46*:

Network

Access Type: Allow secure access from specified IPs and VCNs

Access Control List: Enabled Edit

Mutual TLS (mTLS) Authentication: Required Edit

Figure 7.46 – ACL for an autonomous database

Click on the **Edit** button to configure the ACL settings. By clicking on the **Edit** button, we will get a window as shown in *Figure 7.47*:

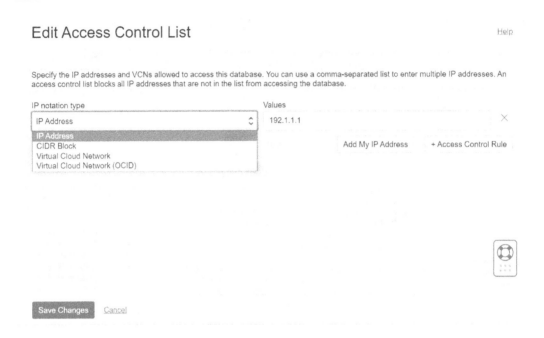

Figure 7.47 – Editing an ACL

The following options are available on the **Edit Access Control List** page:

- **IP Address**: Public IP addresses of the clients that are allowed to connect to the autonomous database. It could be an IP for an on-premises system or OCI VM IP or a third-party cloud VM IP. The **Add My IP Address** button will identify the currently connected system IP address and place the value.

- **CIDR Block**: Public CIDR block of the clients that are visible on the public internet and need access to an autonomous database.

- **Virtual Cloud Network**: A particular VCN available in that region can be chosen to make connectivity with the database. It means only the VMs and applications that are part of the VCN can connect to the autonomous database.

- **Virtual Cloud Network (OCID)**: It's similar to the previous option. Here, we can directly feed the OCID of the VCN instead of choosing from the drop-down list.

After choosing the appropriate option, click on the **Save Changes** button. The changes will be reflected on the details page:

Network

Access Type: Allow secure access from specified IPs and VCNs

Access Control List: Enabled Edit

Mutual TLS (mTLS) Authentication: Required Edit

Figure 7.48 – ACL of the autonomous database

We can see the ACL is enabled on the details page, as shown in *Figure 7.48*.

SQL*Net Connection

All client connections to the autonomous database must use TCP/IP and SSL. This has been configured via the SSL_SERVER_DN_MATCH parameter in the sqlnet.ora file. The sqlnet.ora file is available in the wallet_<servicename>.zip file, which is downloaded through the autonomous database client credential, as shown in *Figure 7.49*:

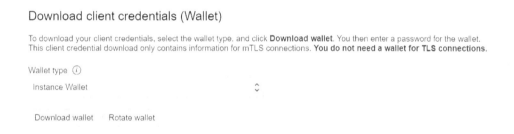

Download client credentials (Wallet)

To download your client credentials, select the wallet type, and click **Download wallet**. You then enter a password for the wallet. This client credential download only contains information for mTLS connections. **You do not need a wallet for TLS connections.**

Wallet type (i)

Instance Wallet

Download wallet   Rotate wallet

Figure 7.49 – Downloading a wallet

The SSL_SERVER_DN_MATCH parameter enforces that the given DN (distinguished name) for the database server matches its service name. Setting the parameter to yes enforces the match verifications and SSL ensures that the certificate is from the database server.

Suppose this parameter is set to no – then, we get the following error:

```
ORA-28860: Fatal SSL error
```

SQL command restrictions

To avoid security violations, certain SQL commands are restricted in the autonomous database. An autonomous database is a self-driven and self-secured database. Hence, it will restrict commands that will affect its security functions. For example, encryption is mandatory in an autonomous database. Hence, SQL commands that close the wallet are not allowed. Tablespaces are managed by an autonomous database. Deletion or addition of tablespaces is not allowed. You will get an error as follows if you try to add or delete tablespaces:

```
SQL> create tablespace users;
Error starting at line : 1 in command -
create tablespace users
Error report -
ORA-01031: insufficient privileges


SQL> Drop tablespace data including contents;
Error starting at line : 6 in command -
Drop tablespace data including contents
Error report -
ORA-01031: insufficient privileges


SQL> administer key management set keystore close;
Error starting at line : 3 in command -
administer key management set keystore close
Error report -
ORA-01031: insufficient privileges
01031. 00000 -  "insufficient privileges"
```

We are also not allowed to close the wallet, as shown here.

Automatic security patch update

Oracle Database gets security patches every quarter. These patches have security bug fixes. In the autonomous database, the security patches are applied automatically every quarter without application downtime. This makes the database secure and not vulnerable to security attacks.

A few environments require a one-off patch to be applied to fix a few bugs seen in that environment. These bugs are specific to the environment and are not generic ones. When the environment migrates to an autonomous database, the patches need to be applied there as well. In an autonomous environment, one-off patches are applied by the support team. We need to raise a support request with patch details and the patch application process will be taken care of by the Oracle team. In case a maintenance window is required for applying the patch, then the customer will be informed in advance.

The Autonomous Dedicated environment has released a new capability for one-off patches. Here, the Oracle engineering team applies the one-off patch immediately rather than waiting for the quarterly planned maintenance window. The default schedule is set in the off-hours of the upcoming weekend with a minimum of 72 hours from the time the patch is available. Off-hours are calculated by Autonomous Database based on database usage. This gives customers the ability to change their schedule to any other suitable time. The **Patch now** option is also provided to apply the patch immediately.

Audit logs

All Autonomous Database operations (start, stop, create, delete, list, get, etc.) are audited by the OCI Audit service. The operations may have been performed by the user or any other utilities such as Cloud Guard or Optimizer. The operations may also have been performed either in the Cloud console or using an API. All activities will be recorded by the OCI Audit service.

In the OCI console, select the **Audit** service under **Identity & Security**. We can see **Audit Events**, as shown in *Figure 7.50*. Choose `Autonomous` as a keyword to list only events related to the autonomous database:

Figure 7.50 – Audit events

We can find the event name, the user who performed the action, at which service it has been performed, the kind of action that has been performed, and the response to that action.

Summary

This chapter explained the security aspects provided by OCI for autonomous databases, in addition to the standard database security features such as Data Redaction and Database Vault. The chapter provided detailed information about Data Safe, which is a free tool for analyzing databases and discovering and masking sensitive data for protection. Having covered security, we have concluded all the topics on Autonomous Database.

FAQs

1. In an autonomous database, can we manage our own encryption keys?

Yes, Autonomous Database allows customer-managed encryption keys. For this, the user needs to create an OCI vault and keys. These have to be rotated by the user at regular intervals.

2. Is Data Redaction available with Oracle Autonomous Database?

Yes, Data Redaction and Data Vault are available with Autonomous Database. All standard database security features are also available with Autonomous Database.

3. I am planning to migrate from an on-premises database to Autonomous Database (the ATP variant). Can Data Safe analyze my on-premises database and recommend safety measures before migrating the database?

Yes. Data Safe can analyze on-premises databases. The database will come under the external database. The database should be reachable by OCI.

4. I would like to migrate data to Autonomous Database using the Data Pump method. Since Autonomous Database doesn't provide server access, where can I place the dump files?

Upload the dump files into an Object Storage bucket and create a database directory in Autonomous Database for the storage location. Autonomous Database can access this storage location to import the dump.

Index

Packtpub.com

Subscribe to our online digital library for full access to over 7,000 books and videos, as well as industry leading tools to help you plan your personal development and advance your career. For more information, please visit our website.

Why subscribe?

- Spend less time learning and more time coding with practical eBooks and Videos from over 4,000 industry professionals

- Improve your learning with Skill Plans built especially for you

- Get a free eBook or video every month

- Fully searchable for easy access to vital information

- Copy and paste, print, and bookmark content

Did you know that Packt offers eBook versions of every book published, with PDF and ePub files available? You can upgrade to the eBook version at packtpub.com and as a print book customer, you are entitled to a discount on the eBook copy. Get in touch with us at customercare@packtpub.com for more details.

At www.packtpub.com, you can also read a collection of free technical articles, sign up for a range of free newsletters, and receive exclusive discounts and offers on Packt books and eBooks.

Other Books You May Enjoy

If you enjoyed this book, you may be interested in these other books by Packt:

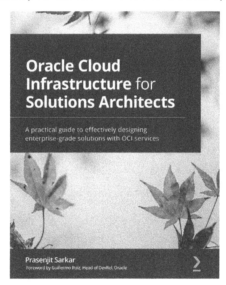

Oracle Cloud Infrastructure for Solutions Architects

Prasenjit Sarkar

ISBN: 9781800566460

- Become well-versed with the building blocks of OCI Gen 2.0 Cloud
- Control access to your cloud resources using IAM components
- Manage and operate various compute instances
- Tune and configure various storage options for your apps
- Develop applications on OCI using OCI Registry (OCIR), Cloud Shell, OCI
- Container Engine for Kubernetes (OKE), and Service Mesh
- Discover ways to use object-relational mapping (ORM) to create infrastructure blocks using Terraform code

Effortless App Development with Oracle Visual Builder

Ankur Jain

ISBN: 9781800569805

- Get started with VB and explore its architecture and basic building blocks
- Gain a clear understanding of business objects and learn how to manage them
- Create service connections to connect to the external API and Oracle SaaS
- Build web and mobile apps and run them on various devices
- Develop Oracle Cloud and non-Oracle SaaS app extensions
- Get to grips with data and application security using practical examples
- Explore best practices along with troubleshooting and debugging mechanisms
- Connect your VB application with VBS for application versioning using Git

Packt is searching for authors like you

If you're interested in becoming an author for Packt, please visit `authors.packtpub.com` and apply today. We have worked with thousands of developers and tech professionals, just like you, to help them share their insight with the global tech community. You can make a general application, apply for a specific hot topic that we are recruiting an author for, or submit your own idea.

Share Your Thoughts

Now you've finished *Oracle Autonomous Database in Enterprise Architecture*, we'd love to hear your thoughts! Scan the QR code below to go straight to the Amazon review page for this book and share your feedback or leave a review on the site that you purchased it from.

`https://packt.link/r/1801072248`

Your review is important to us and the tech community and will help us make sure we're delivering excellent quality content.

Download a free PDF copy of this book

Thanks for purchasing this book!

Do you like to read on the go but are unable to carry your print books everywhere?

Is your eBook purchase not compatible with the device of your choice?

Don't worry, now with every Packt book you get a DRM-free PDF version of that book at no cost.

Read anywhere, any place, on any device. Search, copy, and paste code from your favorite technical books directly into your application.

The perks don't stop there, you can get exclusive access to discounts, newsletters, and great free content in your inbox daily

Follow these simple steps to get the benefits:

1. Scan the QR code or visit the link below

https://packt.link/free-ebook/9781801072243

2. Submit your proof of purchase
3. That's it! We'll send your free PDF and other benefits to your email directly

www.ingramcontent.com/pod-product-compliance
Lightning Source LLC
Chambersburg PA
CBHW060519060326
40690CB00017B/3327